The Effects of Project Labor Agreements on the Production of Affordable Housing

Evidence from Proposition HHH

JASON M. WARD

Sponsored by the Lowy Family Group

RAND SOCIAL AND ECONOMIC WELL-BEING

For more information on this publication, visit **www.rand.org/t/RRA1362-1**.

About RAND

The RAND Corporation is a research organization that develops solutions to public policy challenges to help make communities throughout the world safer and more secure, healthier and more prosperous. RAND is nonprofit, nonpartisan, and committed to the public interest. To learn more about RAND, visit www.rand.org.

Research Integrity

Our mission to help improve policy and decisionmaking through research and analysis is enabled through our core values of quality and objectivity and our unwavering commitment to the highest level of integrity and ethical behavior. To help ensure our research and analysis are rigorous, objective, and nonpartisan, we subject our research publications to a robust and exacting quality-assurance process; avoid both the appearance and reality of financial and other conflicts of interest through staff training, project screening, and a policy of mandatory disclosure; and pursue transparency in our research engagements through our commitment to the open publication of our research findings and recommendations, disclosure of the source of funding of published research, and policies to ensure intellectual independence. For more information, visit www.rand.org/about/principles.

RAND's publications do not necessarily reflect the opinions of its research clients and sponsors.

About This Report

This report presents results from a study focusing on how attaching a project labor agreement—a mandatory contract requiring the use of a primarily union construction workforce and regulating other key aspects of staffing and the utilization of labor on a job site—to Proposition HHH, a large-scale fiscal initiative meant to spur the production of permanent supportive housing projects in Los Angeles, affected both the amount of housing produced and the cost of producing it. The goal of this empirical study is to contribute to a better understanding of the trade-offs involved in combining housing and labor policies. Consistent with the RAND Corporation's mission to provide rigorous, objective, nonpartisan research and analysis, all the data and code related to this project is being made publicly available so that interested researchers may replicate and further explore the results presented herein.

This research was conducted by the Center for Housing and Homelessness in Los Angeles (CHHLA), part of the Community Health and Environmental Policy Program within RAND's Social and Economic Well-Being (SEW) division. The Center for Housing and Homelessness in Los Angeles is focused on providing policymakers and stakeholders with timely research and analysis addressing the dual crises of housing affordability and homelessness in the Los Angeles region and beyond. For more information, visit www.rand.org/chhla.

RAND Social and Economic Well-Being is a division of the RAND Corporation that seeks to actively improve the health and social and economic well-being of populations and communities throughout the world. This research was conducted in the Community Health and Environmental Policy Program within RAND Social and Economic Well-Being. The program focuses on such topics as infrastructure, science and technology, community design, community health promotion, migration and population dynamics, transportation, energy, and climate and the environment, as well as other policy concerns that are influenced by the natural and built environment, technology, and community organizations and institutions that affect well-being. For more information, email chep@rand.org.

Acknowledgments

I thank Alan Greenlee of the Southern California Association of Non-Profit Housing, Julian Gross of the Renne Public Law Group, and numerous other individuals who preferred to remain anonymous for helpful comments and feedback on the background and history of Proposition HHH and project labor agreements. Thanks also to Steven Sharp of Urbanize LA for the cover photo of the HHH-funded Missouri Place housing project.

Thanks to my RAND colleagues Lisa Abraham for providing internal peer review, Daniel Schwam for programming review, Grace Gahlon and Kelsey O'Hollaren for research assistance, and Tiffany Hruby for manuscript preparation. Thanks to Carolina Reid at the Terner Center for Housing Innovation and Robert Santillano at the California Policy Lab for external peer review of this report. Finally, thanks to my CHHLA colleague Sarah Hunter for general support and guidance throughout this project.

Funding

Funding for this research was provided by the Lowy family, whose generous gift supported the establishment of the RAND Center on Housing and Homelessness in Los Angeles in 2020.

Summary

In 2016, Los Angeles voters passed Proposition HHH, directing $1.2 billion in bond funds to support the construction of a pipeline of housing for people experiencing homelessness. Extensive publicity during the campaign for the ballot measure suggested the funding would support the creation of up to 10,000 housing units. However, at present, virtually all funding has been committed, and a total of around 7,300 units of housing are in the pipeline. The failure to meet this more ambitious original target has been attributed, at least in part, to significantly higher-than-expected construction costs, which have averaged around $560,000 per unit to date, an amount that exceeded estimates used during the campaign by around 40 percent. This report provides an empirical assessment of the effects of one candidate mechanism for increased costs, a project labor agreement (PLA) governing HHH-funded projects adopted by the Los Angeles city council approximately 18 months after the passage of the ballot initiative. A PLA is a pre-bid contract governing construction on a project or set of projects agreed to between the funding entity (typically a government entity in the case of public works PLAs) and area construction unions. In order for a contractor to win a contract for a covered project, the contractor must become a signatory to the PLA. A PLA specifies a variety of rules concerning hiring authority, worker ratios (both union and nonunion workers and journey- and apprentice-level union workers), clauses guaranteeing no strikes or lockouts, grievance and arbitration procedures, and in some cases (the HHH PLA is such a case), targeted hiring provisions requiring the employment of local and/or historically disadvantaged workers on the project. Critics of PLAs suggest that these agreements directly increase costs by disincentivizing bidding on projects by nonunion contractors and reducing contractor flexibility over the composition and utilization of the workforce. Advocates of PLAs argue that PLAs lower costs through timely completion of projects and increase competitiveness by leveling the playing field for union contractors with respect to wages, benefits, and other factors. At present, these issues remain disputed and highly contentious, due at least in part to significant limitations in the research designs used in most prior studies on these questions.

The Setting and Design of This Study

The HHH PLA was not part of the ballot initiative that voters passed in 2016. It was added by the Los Angeles city council more than a year later. The primary motivation cited by the council in adding this feature was to ensure that the significant public spending represented by HHH was deployed in a way that supported the employment of local residents and, particularly, residents from disadvantaged backgrounds. This goal was operationalized by including a set of targeted hiring provisions in the PLA that specified goals over numbers of local residents and disadvantaged workers that should be hired on HHH-funded projects. As regards another potential motivation for the use of a PLA, ensuring a fair wage for workers, it is important to note that all developers of HHH-funded projects are required to pay workers "prevailing" (union-level) wages that are specified annually by the state of California.

The HHH PLA is notably different from most public works PLAs, which are associated with public works initiatives such as the building of a new school that has been planned, designed, and put out for bid among contractors. In such cases, any contractor wishing to build the school must become a signatory to the PLA and must build the specified building. The PLA may influence the cost of the buildings but not the size, location, and so on.

Proposition HHH, by contrast, is a *funding* program that was intended to fund scores of projects independently proposed by affordable housing developers. The PLA associated with HHH applied to a given project only if it comprised 65 housing units or more. This combination of a prespecified size-based threshold and the decentralized manner in which projects were conceived and proposed allowed the total number of housing units built to be directly influenced by the PLA. Setting a threshold that only covered larger developments rendered these projects potentially more costly, working directly against other incentives (e.g., interest rates) in HHH meant to incentivize larger projects.

This setting provides a rare opportunity to generate convincing evidence on the effects of PLAs on the development of affordable housing. I use two specific approaches to estimate these effects. First, I estimate how the PLA affected the size of proposed housing projects funded through HHH by comparing the difference in the shares of HHH-funded projects above and below the 65-unit PLA threshold with the difference in these same shares among a sample of similar non–HHH-funded projects. Second, I use a regression model that estimates the effect of the PLA on construction costs by comparing the size of cost discontinuities at the 65-unit threshold between these two samples (while controlling for a variety of important factors that also affect project costs). Finally, in a simulation exercise that combines these two approaches, I estimate how many housing units might have been produced in the absence of the PLA.

Key Findings

Four key findings result from this analysis:

1. The evidence strongly suggests that developers responded to the PLA by disproportionately proposing housing projects that fell below the 65-unit threshold. Figure S.1 shows the frequency distribution of projects according to the number of housing units they contain. As can be seen, there is a dramatic decline in the number of proposed projects as the number of units crosses the PLA threshold. In total, 22 of the 98 total new construction projects in the data are in the narrow range of 60 to 64 units. In contrast, there was one proposed project falling between 65 and 69 units.

Figure S.1. Frequency Distribution of Project Sizes for HHH-Funded Projects

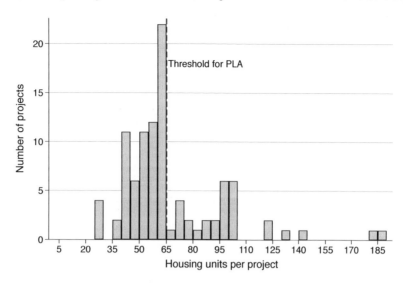

SOURCE: Author calculations using city of Los Angeles, TCAC, and CDLAC data.

2. Comparing this distribution of project size with a sample of non–HHH-funded affordable/supportive housing projects provides strong evidence that this discontinuity in project sizes was caused by developer response to the presence of the PLA. Figure S.2 presents a comparison of the shares of projects by size for HHH-funded projects and non–HHH-funded projects. While projects with 50 to 64 units make up more than 45 percent of the HHH sample, such projects make up less than 10 percent of the non-HHH sample. Meanwhile, relative to the corresponding HHH shares, the share of non-HHH projects with between 65 to 79 units is around 100 percent larger and the share with between 80 and 94 units is approximately 250 percent larger.

Figure S.2. Distribution of Project-Size Shares by Funding Source

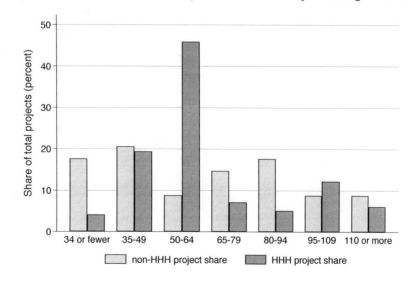

SOURCE: Author calculations using city of Los Angeles, TCAC, and CDLAC data.

3. After accounting for a variety of important characteristics of individual projects, I estimate that per unit construction costs were approximately $43,000 higher for projects covered by the PLA. This amounts to a 14.5-percent increase in construction costs.
4. In a simulation exercise I estimate that, in the absence of the HHH PLA, a combination of developers building larger projects and lower costs facilitating the funding of more projects would have resulted in approximately 800 more units of housing, or an amount representing around 11 percent of the total of 7,305 housing units in the actual HHH pipeline today.

Policy Considerations

The findings indicate that the inclusion of PLAs and similar labor regulations to funding programs such as HHH is likely to influence the primary housing production goals of such policies. In the case of the Proposition HHH, the use of a PLA with a housing unit-based threshold reduced the total housing produced through two channels, reducing the total housing units in a significant number of funded projects and increasing the cost of each housing unit in projects covered by the agreement.

It is unclear why developers responded so strongly to the presence of the PLA. The developer community building HHH projects is primarily composed of nonprofit, mission-driven organizations that are generally allowed to charge a capped developer fee related directly to project costs. These factors mitigate against the likelihood that concerns over foregone profits related to higher costs would create incentives similar to what might be expected of developers of market-rate housing.

Developer concerns about the PLA adding uncertainty over costs and timelines may have been an important factor. Deeply subsidized affordable housing projects already face considerable uncertainty related to community opposition, assembling the necessary funding, and uncertain timelines for regulatory approvals. The PLA may have represented one source of uncertainty that was avoidable through the choice of a smaller project size.

Perhaps more concretely, potential cost drivers related to size—including the HHH PLA but also, for example, Community Development Block Grant funding that requires the payment of prevailing (union-level) wages if a developer pursues a housing project with eight or more housing units—may influence project size directly by raising costs such that a potential project will no longer be financially feasible because of both absolute limits on amounts available from various funding sources (e.g., maximums on per-project funding using low-income housing tax credits) and constraints on their use (e.g., funding through certain state of California programs cannot be "stacked" or used together for a single project).

Regarding the city council's stated motivation in adopting the PLA—that HHH funding provide quality employment opportunities to local residents and disadvantaged workers—one alternative to using a PLA would be to rely on a "first source hiring" ordinance, such as one already in use by the city of Los Angeles for a variety of city contracts. First source requirements are currently used for housing projects in the nearby municipality of Pasadena, where local hiring is ensured by direct, enforceable requirements (there is no explicit enforcement mechanism for failing to meet these goals in the HHH PLA).

If, instead, a primary policy goal is to ensure that publicly funded housing is built primarily or exclusively using a unionized workforce, then a more transparent policy would be to include this stipulation in similar future initiatives and legislation, rather than adding it as a post hoc requirement. More generally, increased transparency around the trade-offs involved in combining housing production policies with restrictive labor regulations may help to set realistic expectations, avoiding the erosion of public and policymaker support that has befallen Proposition HHH.

Contents

Figures and Tables

Figures

Tables

1. Introduction

In recent decades, Los Angeles has become a national focal point for the lack of affordable housing that is creating housing instability for millions in booming metro areas around the United States (Woetzel et al., 2019). This affordability crisis has been associated with a substantial rise in homelessness that has become the most pressing local issue for a majority of Angelenos (Byrne, Henwood, and Orlando, 2021). In response, an unprecedented local ballot initiative known as Proposition HHH was put before voters in the city of Los Angeles in 2016. The initiative proposed to raise $1.2 billion through a property tax levy in order to fund a massive expansion of the stock of permanent supportive housing (PSH) in Los Angeles.[1] While the language of the ballot initiative did not propose a specific goal for the amount of housing to be produced, it was widely characterized in city outreach and in the media as a measure that would create up to 10,000 units of PSH/affordable housing (Holland, 2016; *Los Angeles Times* Editorial Department, 2016). In an early Q&A document from the city, the goal of 8,000 to 10,000 units was characterized as "conservative" (City of Los Angeles, undated). It was projected that 7,000 of these affordable units would specifically be PSH (Fiore et al., 2019). To put this goal in context, an increase of this size in the city's supportive housing unit portfolio in 2016 represented around a 40-percent increase in the approximately 16,000 total units of PSH available in the entire *county* of Los Angeles (Conrad N. Hilton Foundation, 2018). In November 2016, the measure passed overwhelmingly, with 76 percent of voters supporting it (Holland and Smith, 2016).

The High Costs of HHH-Funded Housing Projects

As of May 2021, the city's HHH data portal reported a total of 7,305 total housing units (5,760 PSH units) in the pipeline with $973 million (81 percent) of the total funding committed, though other sources report that more than 97 percent of the funding has been conditionally awarded (Galperin, 2019; *Los Angeles Times* Editorial Board, 2021). This pipeline of funded projects, when completed, will provide an unprecedented increase in the total number of supportive housing units in Los Angeles, an achievement that should be recognized and celebrated. But, five years on, the nature of coverage and debate around HHH suggests that this achievement has been overshadowed by the impression that HHH has failed to deliver on its

---

[1] "Permanent supportive housing" (or sometimes simply "supportive housing") is a broad term used to indicate housing units intended as permanent residences that are paired with case management and service provision for, as examples, mental health, addiction, or other special needs of chronically homeless individuals (National Alliance to End Homelessness, 2020).

ambitious initial promise. Much of the blame for the shortfall in housing units has been attributed to the fact that the average estimated cost of awarded projects is much higher than per unit cost estimates used during the HHH campaign. These estimates ranged from $350,000 to $420,000 per PSH unit (Holland, 2016; Smith, 2016). Estimated costs of awarded projects have been significantly higher, averaging around $560,000 per unit, with outlier projects reaching more than $700,000 per unit (Galperin, 2019; Sharp, 2020).[2]

Such high costs have led to calls for reconsidering how HHH funding is being spent. In late 2019, the Los Angeles city controller released a report calling into question the wisdom of using virtually all HHH funds to build a relatively small stock (relative to immediate need) of new construction housing units and suggested exploring reallocating some of the committed but as-yet-unspent funding for use on alternative arrangements, such as temporary shelters that could address street homelessness in months rather than years (Galperin, 2019). City council member Kevin DeLeon has recently echoed these calls (Oreskes, 2021).

Cost Components of HHH Projects

In order to think about potential drivers of these high costs and to motivate the focus on construction costs, it is helpful to briefly summarize some housing development cost basics. Costs in affordable housing development are commonly grouped into three major categories: "hard" costs (analogous to construction costs, the term we will retain in this report for clarity), land costs, and "soft" costs.[3] Construction costs are labor, materials, and various ancillary costs related to site preparation and construction. Land costs are, as might be expected, the costs of land acquisition. Soft costs is a residual category representing all other project-related costs. Significant components of soft costs typically include financing costs, architectural and engineering fees, and permitting and impact fees. In this report, costs that are closely but indirectly related to construction costs, including developer fees and construction cost contingencies, are included in the soft cost category.

Figure 1.1 shows the shares of these three cost categories in percent terms for the analytic sample of HHH projects.[4] Construction costs make up just over 60 percent of total per unit costs among these projects. Land costs represent less than 10 percent of total costs. The small size of this share in the high-cost environment of Los Angeles is driven by the fact that a number of

---

[2] Most projects only have estimated costs at this time since they are either in the pre-construction or construction phases, but the seven completed projects as of March 2021 had actual per unit costs that averaged $528,187—a figure that is the author's calculation using data from the Los Angeles Housing and Community Investment Department, received pursuant to a California Public Records Act request.

[3] A fourth category, "conversion costs," is sometimes used as well and refers to costs incurred at the completion of a project, such as title fees and operating reserves (Raetz et al., 2020). For this report, these costs are included as soft costs.

[4] This sample and a sample of comparison non–HHH-funded projects are described in detail in Chapter Two.

developments use donated city or county land. The remaining 30 percent of total costs are soft costs. These shares are broadly consistent with other recent affordable housing developments in California (see Raetz et al., 2020).

Figure 1.1. Cost Shares of HHH Projects

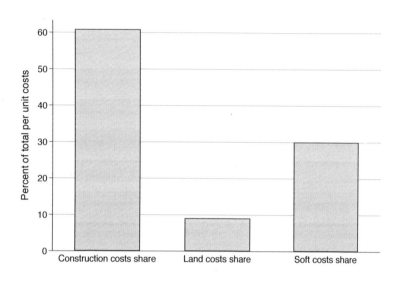

SOURCE: Author calculations from city of Los Angeles, TCAC, and CDLAC data.

Motivation for this report's focus on construction costs is presented in Figure 1.2, which displays these same cost categories in dollar terms for both the HHH sample and a companion sample of non–HHH-funded PSH projects that closely overlap the HHH projects in time and are highly comparable in all other substantive ways of which I am aware (these two samples are discussed in more detail in Chapter Three). As can be seen, these projects have nearly identical land costs and generally comparable soft costs. However, the construction costs of HHH-funded projects are, on average, $81,000 (31 percent) higher per unit than the non–HHH-funded projects. This substantial cost difference can explain a significant portion of the discrepancy between the projected per unit costs used during the HHH campaign and the realized cost estimates of developments funded by the initiative. As I demonstrate below, a significant portion of this average cost difference appears related to the HHH PLA. However, understanding what other factors figured into this larger overall cost difference is an important area for future research.

Figure 1.2. Average per Unit Costs by Category for HHH and Non-HHH Projects

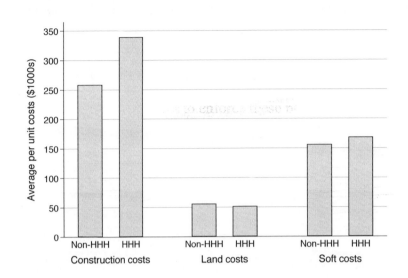

SOURCE: Author calculations from city of Los Angeles, TCAC, and CDLAC data.

Existing Research on Cost Drivers of HHH Projects and Other Affordable Housing

Existing analysis of cost drivers among HHH projects has focused primarily on non–construction-related channels. A recent report from Abt Associates pointed to fiscal factors such as the 2017 federal tax reform bill that diminished the value of the low-income housing tax credit (LIHTC) and an initial lack of access to funding from the state's No Place Like Home program related to a lawsuit over the legality of using these funds to construct new housing (Fiore et al., 2019). Media reports have highlighted legal challenges to the city's 2018 permanent supportive housing ordinance and the use of so-called pocket vetoes by city council members in creating legal barriers to moving forward with projects under HHH (Alpert Reyes, 2018; Smith, 2018).

I am unaware of any studies focusing specifically on construction costs of HHH projects, but a 2014 report from multiple state agencies and a pair of recent studies from the Terner Center for Housing Innovation at the University of California, Berkeley have more broadly explored the role of construction costs in the growth of affordable housing costs in the state (California Department of Housing and Community Development et al., 2014; Raetz et al., 2020; Reid, 2020). These studies point to increases in both material and labor costs as significant cost drivers in recent years, finding that, for example, construction costs of multifamily housing in California rose by around 25 percent between 2009 and 2018 and that affordable projects cost more to build on average than market-rate projects due to a mix of factors, including higher development costs and differences in average unit size.

An important additional contribution of these studies is estimating the association between prevailing wage (PW) laws and construction costs on the production of affordable housing in

California. Prevailing wage laws apply to most affordable housing projects using public funding in California and require that workers building these projects be paid the prevailing union wage for their job along with certain benefits or additional wages equivalent to the value of these benefits. This requirement is less typical in market-rate developments.[5] These recent Terner Center reports (Raetz et al., 2020; Reid, 2020) find that PW laws increase construction costs by between 11 and 16 percent. Note that all HHH projects and the majority of the non–HHH-funded projects used in this analysis are subject to prevailing wage laws. For more on prevailing wage laws and associated research, see Duncan and Ormiston (2019).

This report focuses attention on one particular aspect of construction costs for HHH-funded projects: a project labor agreement (PLA) regulating several important aspects of the staffing and utilization of labor in the construction process. This policy is of specific interest for multiple reasons, including the following:

- Despite the fact that it had potentially important ramifications for the effectiveness of Proposition HHH, the PLA was not included in the language of the ballot initiative.[6]
- During the debate between developers, trade union representatives, and the city over the structure of the PLA, none of the criteria suggested by the developer community, including a higher unit threshold (75), a less restrictive definition of core workers, an exemption for subcontracts that receive fewer than three bids, and a shorter initial term (two years versus five) for the agreement to apply, were incorporated into the final language of the agreement, suggesting that the process discounted the impact that developer behavior could have on the subsequent outcomes of HHH funding.
- The primary motivation cited by the city council for adopting a PLA—a desire to expand career pathways into the construction trades for local residents and targeted groups—is a relatively small and voluntary component of the full suite of requirements in the PLA, suggesting either that the potential overall effects of the agreement on housing production were not well understood or that the motivation for the adoption of the PLA was not effectively communicated at the time of its adoption.

Despite a vigorous and ongoing debate about the effects of PLAs on costs and competitiveness (as documented in the next chapter), the effects of this agreement have received no significant attention in the debate over the high costs of HHH-funded housing. This report attempts to fill important knowledge gaps around the determinants of the unexpectedly high costs of projects built using HHH funding and to further the debate on the effects of PLAs more generally.

---

[5] There are exceptions related to specific legislation, such as Los Angeles's Measure JJJ, which requires that market-rate projects seeking general plan amendments or zoning changes pay all workers PW (City of Los Angeles Department of Public Works, 2021).

[6] This is in contrast to Measure JJJ, where the requirement that developers seeking amendments or zoning changes for projects pay prevailing wages was included in the language of the ballot initiative. JJJ was passed at the same time as Proposition HHH.

The HHH Project Labor Agreement

A PLA is a pre-bid contract between the entity funding a project and area construction unions that specifies a variety of rules, including steering all hiring through union halls, limiting the number of nonunion "core" workers a contractor may use on a project, specifying worker ratios (between apprentice and journey-level workers), and guaranteeing no strikes or lockouts, and that has grievance and arbitration procedures to enforce these provisions. Once a PLA is in place, any contractor wishing to bid on/build associated projects must become a signatory to this agreement. Historically, PLAs have been most common in the private sector, but they are also widely used in the public sector for large-scale, government-funded projects (General Accounting Office, 1991). In recent decades, PLAs governing public works projects (such as the HHH PLA) have increasingly included language prescribing various levels of local hiring or the hiring of traditionally underrepresented or disadvantaged workers who have completed union apprenticeship programs (Figueroa, Grabelsky, and Lamare, 2011).

The effect of PLAs on construction projects is a hotly disputed issue, with critics suggesting that PLAs substantially increase costs and reduce competitiveness by disincentivizing nonunion contractors from bidding on relevant projects and advocates countering that PLAs have no distinguishable effects on project costs or the number of contractors that bid on covered projects (Brothers, 2020; Brubeck, 2021; Calandro, 2021; ENR California, 2011; Martindale, 2013). Existing research on these questions is broadly characterized by weak study designs primarily related to a lack of comparability between PLA and non-PLA projects. Additionally, much of this research is supported or directly sponsored by organizations with strong ex ante positions on the use of PLAs.

The PLA governing projects that receive HHH funding was not part of the original ballot initiative. It was adopted by the city council approximately 18 months after the passage of Proposition HHH. The primary motivation provided by the city council for implementing a PLA for HHH-funded projects was to increase career pathways into construction work for local residents and disadvantaged workers. The council's motion of May 2017 states:

> With the approval of Proposition HHH by the voters in November 2016, the City of Los Angeles is embarking on a $1.2 billion . . . housing and facilities construction program with the goal of creating 10,000 permanent supportive housing units to address citywide homelessness and the homeless housing shortage. As part of the implementation of this measure, the City should create a policy that reinvests bond dollars into our local neighborhoods and residents by training and employing them as often as possible on funded projects, while maintaining the unit goal of Proposition HHH. (Cedillo et al., 2017)

It is important to note that local and targeted hiring ordinances are common tools to achieve such goals and that the city of Los Angeles has had such programs for some time—notably, a first source hiring ordinance that applies to city contractors (City of Los Angeles, 2016). In the case of HHH, however, it was decided to use a PLA with these hiring goals operationalized through the inclusion of a targeted hiring provision (THP).

For the HHH PLA, the THPs specified that 30 percent of work hours must be performed by individuals from a tiered system of geographic proximity, and 10 percent of hours (one-third of the 30 percent) must be performed by disadvantaged (or "transitional") workers from within these residential areas. Among residents, two tiers of transitional workers are defined. The first is veterans and those with a history of criminal justice involvement. If individuals meeting these criteria cannot be sourced, a second set of characteristics can be used that includes having two or more of the following barriers to employment: household income of less than 50 percent of the county median, receiving public assistance, educational attainment below the high school level, being a custodial single parent, experiencing long-term unemployment, or being emancipated from the foster care system. One additional criterion in this second tier that is not clearly associated with social disadvantage is also included: being a current union apprentice who has satisfied less than 15 percent of the apprenticeship hours required to graduate to the journey level (City of Los Angeles, 2018).

However, the goals in the resulting PLA (and similar PLAs used in Los Angeles and elsewhere) for the hiring of local residents and targeted subpopulations of residents lack a clear enforcement mechanism, using what is sometimes characterized as "good faith" language (Luster et al., 2010). The limited literature on these community workforce provisions (summarized in Chapter Two and in Table C.1) suggests that failing to achieve these goals is relatively common.

Typical public works PLAs—such as those governing construction projects in the Los Angeles Unified School District, the Los Angeles Community College District, and Los Angeles Metro—have a cost-based threshold governing what projects fall under the agreement, with projects above some maximum amount being subject to a PLA (Los Angeles Community College District, 2001; Los Angeles Unified School District, 2003; Los Angeles County Metropolitan Transportation Authority, 2017). The HHH PLA's threshold, however, was unusual in that it used the number of housing units proposed (65 or more) as the threshold for a project to require the use of the PLA. A report from the city's Bureau of Contract Administration summarizing the process of drafting the PLA implies that this threshold was chosen based on an expectation that the 65-unit threshold would lead to around half of total HHH projects falling under the PLA (Reamer, 2017).[7]

Two aspects of the HHH PLA are remarkable relative to how PLAs are typically used in public works projects. The first is that the HHH PLA is an example of a government entity requiring the use of a PLA on privately operated construction projects as a condition for the project receiving partial public funding. This is quite different from typical PLAs used in cases when the project itself is a publicly owned piece of infrastructure. Second, unlike most PLAs that accompany a construction project where the characteristics of the infrastructure to be built are

---

[7] The HHH PLA also has a $5 million total development cost threshold for facilities improvement projects (work that is ancillary to the main buildings containing housing, such as site improvements), but this report focuses on housing projects specifically.

clearly specified in advance, the HHH PLA was attached to an agreement that specified the funding available for building infrastructure, without specifying key features of the infrastructure to be built. The implementation of a PLA in this unusual context and the use of a threshold applying the PLA only to projects with 65 or more housing units left developers with a choice to be subject to the PLA or not based on the number of housing units they proposed to include in a development. Surprisingly, the Bureau of Contract Administration report does not consider at all the possibility that developers might respond to the PLA threshold in such a manner. This possibility is a primary motivation for this report.

The Organization of This Report

The next chapter (Chapter Two) provides a focused review of relevant existing research on PLAs with attention to the plausibility of the research methods as well as the provenance or funding of the research as it relates to organizational support for or opposition to PLAs. Chapter Three discusses the setting and motivation and provides the intuition for the research designs employed. Chapter Four presents evidence on the association between the PLA threshold and the shares of projects by size (number of housing units). Chapter Five presents the design of the cost-analysis model and estimates of the causal effect of the PLA on per unit construction costs. Chapter Six contains the results of a simulation exercise that generates counterfactual estimates of the number of units produced and the cost of producing those units, had HHH funding been provided without the PLA requirement. Chapter Seven concludes the report with a discussion of the policy implications of these findings and provides suggestions for future, related research and policy. Multiple appendixes are also included that provide greater detail on the project data, the regression model used, results from multiple sensitivity tests, and other supporting evidence.

2. Existing Research on Project Labor Agreements

Empirical evidence of the relationship between mandatory public works PLAs and project costs, number of bidders, worksite safety, and the hiring of local and disadvantaged workers has been accruing since the first large-scale public works PLA of the modern era was enacted in 1993 to cover the nearly decade-long cleanup of Boston Harbor (Northrup and Alario, 1998). However, despite the considerable judicial and legislative attention that has been focused on PLAs due to both the financial and political stakes these agreements are associated with, the evidence on their effects remains controversial and relatively scant.

The lack of clarity around how PLAs affect competitiveness, costs, and other aspects of public works projects has two key drivers. The first is the difficulty of finding real-world settings where conditions allow for an apples-to-apples comparison of these outcomes on PLA and non-PLA projects. A 1998 General Accounting Office (GAO) report noted that both "proponents and opponents of the use of PLAs said it would be difficult to compare contractor performance on federal projects with and without PLAs because it is highly unlikely that two such projects could be found that were sufficiently similar in cost, size, scope, and timing" (GAO, 1991).

The second is the fact that most research on PLAs has been conducted by researchers with clear pro- or anti-PLA affiliations. As summarized below, the majority of the research concluding that PLAs increase project costs and reduce competition has been conducted or funded by groups with a strong prior anti-PLA policy position. Relatedly, all the summarized studies finding that PLAs do not increase costs or affect competition have been conducted by researchers supported by groups either directly or indirectly affiliated with labor unions. While these affiliations should not be viewed as prima facie evidence of bias, they may provide context for the deployment of nonstandard research designs or limited sensitivity testing performed on results. Very few studies concerning PLAs have been conducted by groups unaffiliated with a prior position on the effects of these agreements.

In this brief review of existing research on PLAs, I focus on two significant aspects of the debate over the costs and benefits of these agreements:[8]

1. *Project costs and competitiveness.* (These two issues are considered separately in some of the studies reviewed below, but they are highly related conceptually, so I group them together for discussion here.) For projects where the absence of a PLA allows contractors

---

[8] Other issues of contention include the relationship between PLAs and strikes and other work stoppages, improved building quality due to increased use of skilled workers, as well as worker safety. For a full accounting of these issues, interested readers may consult two useful overview articles. One, Northrup and Alario (1998) is written from an anti-PLA point of view. A second one, Lund and Oswald (2001), is written from a pro-PLA view and specifically addresses or rebuts some of the issues raised in the first article.

to bid on jobs using market-rate (i.e., non-prevailing) wages, PLAs directly increase the wage bill on a project by requiring a primarily union workforce. Additionally, in cases where workers would be paid prevailing wages in the absence of a PLA, opponents point to higher labor costs related to what are commonly called "terms and conditions" (various types of additional pay stipulations related to travel, holidays, and overtime) as well as differences in overall staffing levels and worker roles on the job site that affect costs (e.g., nonskilled tasks such as moving materials being performed by apprentice or journey-level union workers rather than lower-paid site helpers) (Northrup and Alario, 1998). The latter claim is disputed by proponents of PLAs who argue that the on-time completion of projects by highly trained workers results in equivalent or lower costs (Lund and Oswald, 2001). Critics further suggest that PLAs decrease competition by limiting the pool of contractors willing or able to bid on a project and that this reduced competition leads to increased project costs through two mechanisms: higher costs among firms that do bid a project with a PLA and markups among this smaller pool of bidders induced by the lack of competition. Proponents of PLAs counter that many nonunion contractors successfully compete for and win projects using PLAs and that both the number of bids for projects using PLAs and bid costs do not generally differ substantively from bids on non-PLA projects.

2. *Hiring of local and disadvantaged workers.* Proponents of PLAs point to the "community workforce provisions" that are a part of many public works PLAs as evidence that PLAs increase local hiring and occupational mobility of disadvantaged workers including women, minorities, veterans, and other groups (Figueroa, Grabelsky, and Lamare, 2011). Critics counter that the lack of representation of these groups among both journey- and apprentice-level union workers, as well as the complex path into and through union apprenticeship programs, may increase barriers to achieving these goals relative to training programs operated by nonunion contractors, where local workers could be hired, trained, and put to work on a shorter time frame (Moran, 2011). These requirements have not historically been an integral component of PLAs but have become more common in recent years (Figueroa, Grabelsky, and Lamare, 2011).

Summary of Findings from Existing Research on PLAs

The following conclusions relating to the issues highlighted above emerge from a review of the existing research on PLAs:

1. The prepondcrance of evidence does suggest that PLAs tend to be associated with higher costs, but the research designs that produce these findings are often highly sensitive to alternate ways of specifying the model and are subject to legitimate criticisms regarding omitted variable bias and a lack of comparability among PLA and non-PLA projects.
2. The evidence on lack of competitiveness, as measured by the number of bids on associated projects, is relatively weak compared with the results around project costs, and these results are subject to the same sorts of criticisms with respect to the research designs employed.
3. Reviewing the research on costs and competitiveness, among the studies conducted or funded by organizations with a policy position on PLAs, the findings are, in every case,

strongly aligned with the policy positions of the organization, an association that does little to contribute to the quality of the debate over these issues.

4. The sparse existing evidence on targeted hiring provisions (THPs) in PLAs suggests that these goals are often not fully met. This may be due, at least in part, to the lack of any substantive accountability mechanisms associated with these goals. Projects under public works PLAs in Los Angeles, with a population of nearly 4 million, failed to meet goals of hiring as few as 30 percent of project workers from among targeted city residents. Missing goals for disadvantaged workers was even more common. Anecdotal evidence from these case studies suggests that the complexity and limited capacity of the union apprenticeship pipeline may play a role in these outcomes.

5. The existing research on THPs under PLAs does not contribute to an understanding of whether PLAs actually *increase* career pipelines into construction work, since none of these studies incorporates measures of these outcomes from non-PLA projects (and some of these studies lack complete data on the attainment of THP goals, even for the PLA under study). This knowledge gap is unlikely to be filled without significant changes to reporting requirements and data collection practices by state and local agencies.

Below I summarize the studies included in my review in more detail. In these summaries, I pay attention to the key limitations of each study design as well as the source of the study (whether there is some financial or organizational affiliation to a pro-PLA organization, an anti-PLA organization, or an entity without a clear association to a policy position on PLAs).

Evidence on the Relationship Between PLAs and Project Costs

At least four studies on the relationship between PLAs and both project costs and the project bidding process have been conducted by the Beacon Hill Institute, a free-market think tank, between 2004 and 2019.[9] These studies have a similar approach of using observational data on the number and amount of project bids and/or actual construction costs for state-specific samples of public works projects (usually schools) and comparing these outcomes according to whether the project had a mandatory PLA. All of this research concludes that PLAs are associated with higher costs. A 2004 study focused on Connecticut estimated that PLAs were associated with an 18-percent increase in school construction costs (Bachman, Haughton, and Tuerck, 2004). A similar study using Massachusetts data found that PLAs were associated with approximately 15 percent higher school construction costs (Bachman and Haughton, 2007), while a 2006 report focused on New York State school projects estimated that PLAs were associated with 25 percent higher construction costs (Bachman and Tuerck, 2006).

A critical weakness of this series of studies is that projects subject to a PLA may differ in important ways—such as geographic differences in construction costs and differences in project

---

[9] The ideological orientation of the institute is not made explicit in its own mission statement, but it is a member of the State Policy Network, an association of think tanks with a mission to "catalyze thriving, durable freedom movements in every state, anchored with high-performing independent think tanks" (State Policy Network, 2021).

characteristics—from non-PLA projects. These confounding influences are unlikely to be addressed by research designs typically employing relatively simple regression models with few controls.

In a 1995 report from the Associated Builders and Contractors, an anti-PLA organization, on the PLA governing the construction of the Roswell Park Cancer Institute in Buffalo, New York, the authors estimated that bid packages for project components that required a PLA were, on average, 10 percent over estimated costs, while bid packages for non-PLA components averaged 13 percent below estimated costs, and that projects without a PLA averaged 21 percent more bids than PLA projects (Associated Builders and Contractors, 1995). But this descriptive comparison sheds little light on how projects may have differed in ways that could influence the number of bids or costs.

In a critique of Bachman and Haughton (2007), Belman et al. (2010) test the sensitivity of the results in Massachusetts. They found that adding a small set of controls (e.g., whether the school was public or private, the number of stories, the presence of a basement) reduced the size of the estimated PLA association by nearly 20 percent. The further addition of a binary indicator variable for whether the project was in Boston, a higher-cost location than elsewhere in Massachusetts that is also the location of most of the PLA projects, reduced the magnitude of the (still positive) association between PLAs and costs by more than half.[10]

A 2010 peer-reviewed report funded in part by the Associated Builders and Contractors estimated the association between PLAs and project costs using 551 California school construction projects from 1996 to 2008 (Vasquez, Glaser, and Bruvold, 2010). Employing a regression model similar to Belman et al. (2010), they found that school projects built under a PLA cost approximately 14 percent more per square foot. But, similar to Bachman and Haughton (2007), while the overall sample comprised projects from around Southern California, the majority of the PLA projects were part of just one district, Los Angeles Unified School District (LAUSD). When the authors independently controlled for projects built in LAUSD, the estimated PLA association decreased by around 70 percent and became statistically insignificant, suggesting that much of the estimated PLA association likely represented aggregated cost factors unique to Los Angeles.

A pro-PLA 2015 working paper that is perhaps most conceptually relevant to the present study focuses on the relationship between PLAs and affordable housing (AH) construction costs in Los Angeles (Philips and Littlehale, 2015). The study used a sample of 130 AH projects developed between 2008 and 2012, including nine AH projects that used a PLA and 121 that did not, and found no associations between the PLA projects and higher costs. However, one of the three analyses was a visual comparison, another was a simple-means comparison of square foot

---

[10] The authors of this study received support from the Center to Protect Worker's Rights, a union-funded advocacy group.

and per unit costs, and the third compared mean cost differences among projects matched using a nearest-neighbor approach, making these results difficult to compare with the results using linear regression models.

The two reports on PLAs and project costs conducted or sponsored by government entities are a 1991 report by the Government Accounting Office and a 2010 report from the New Jersey Department of Labor (DOL). The GAO report (GAO, 1991) was in response to a Senate request to investigate complaints from nonunion contractors about access to construction projects at a large Department of Energy facility in Idaho that adopted a site-specific PLA in the mid-1980s. The GAO found that average wages for projects under the PLA were around 20 percent higher than before the PLA, when only a prevailing wage requirement applied. The New Jersey DOL report focused on state school construction projects from 2007 to 2008 and found that, after controlling for geographic cost differences, project size, and school type, PLA projects had costs per square foot that were approximately 20 percent higher than non-PLA projects (New Jersey Department of Labor, 2010).

Evidence on the Relationship Between PLAs and Competitiveness

Turning to the association between PLAs and bidding on projects, a 2019 report authored by researchers from two free-market research organizations, Beacon Hill and the Washington Policy Center, used data on 125 projects from across Washington State. The authors found that PLAs were associated with around 10 percent fewer bids per project (Bachman, Burke, and Tuerck, 2019). However, the model did not control for the type or locale of the project and did not make clear whether, for instance, PLA projects were located in higher-cost areas such as Seattle.

A 2017 study funded in part by the Marin County Building Trades Council compared bids and bid amounts among 263 PLA and non-PLA California community college construction projects (Philips and Waitzman, 2017). The authors concluded there was no statistically significant difference in the number of bids between PLA and non-PLA projects. For reasons not made entirely clear in the report, the authors used an unconventional analysis that make these findings difficult to directly compare with other analyses.[11]

Evidence on the Relationship Between PLAs and Targeted Hiring

A less common but growing characteristic of public works PLAs are THPs that require various types of local hiring or the hiring of disadvantaged workers of special interest (e.g., women, racial and ethnic minorities, residents of low-income areas, individuals with a history of criminal justice

---

[11] Specifically, the authors' model predicted the number of bids controlling for a quadratic term in the low bid amount as a proxy for project size. No other approaches, such as median or average bid, are explored, nor are more conventional measures of project size, such as the number of stories of the relevant building or estimated project costs.

involvement, or veterans). The rationale behind these provisions is that public works money should support occupational and social mobility for disadvantaged or at-risk workers by creating career pathways to well-paying jobs.

The small amount of existing evidence on how often the goals laid out in THPs are achieved is mixed. It is important to note that there is no evidence that I am aware of comparing local hiring outcomes from PLA projects with THPs and local hiring under any sort of "status quo" discretionary hiring process, either with prevailing or market wages.

Two relatively recent reports provide evidence on whether specified THP goals were met for PLAs in Cleveland, Washington, D.C., and New York City (Figueroa, Grabelsky, and Lamare, 2011) and for three PLAs in the Los Angeles region (Owens-Wilson, 2010).[12] The typical goals in the THPs associated with the PLAs in these studies were related to share of work performed by local residents, share of new hires from among local residents, and both share of work performed and apprentices hired from among disadvantaged workers. Across these two reports, approximately half of the goals set out in the PLAs of focus for which outcome data were available were met. The most common area of failure was in meeting goals around the hiring of disadvantaged workers.

A 2010 report on the construction industry in the greater San Francisco area commissioned by San Francisco's redevelopment agency provides analysis of several THPs that are implemented both as stand-alone policies and as part of a PLA (Luster et al., 2010). Their analysis highlights that the vast majority of these agreements, including all such examples of agreements within a PLA, rely on "good faith" language (often accompanied by compliance monitoring of some sort) to incentivize compliance. The summary table provided by the authors indicates that, across the 11 programs they were able to collect data for on both goals and outcomes, approximately 50 percent of worker/hours goals were met under local hire programs outside of a PLA, while approximately 33 percent of worker/hours goals were met under THPs in a PLA.

Table C.1 presents all the main findings from the studies summarized above that had sufficiently comparable outcomes.

---

[12] Figueroa, Grabelsky, and Lamare (2011) was funded by the American Rights at Work Fund, an advocacy group focused on advancing union organizing. Owens-Wilson (2010) was funded by the Partnership for Working Families, an advocacy organization supported by both philanthropies and labor unions.

3. Setting, Research Design, and Data Used in This Report

Los Angeles and Proposition HHH Offer a High-Quality Research Setting

In contrast with the suboptimal settings typical of the existing research on PLAs (summarized in Chapter Two), the setting for the study reported here is remarkably ideal, both in terms of capturing the causal effects of a PLA on project costs as well as providing causal estimates of how a PLA affects developer incentives that have not, to my knowledge, been explored in any previous study. The key characteristics of the HHH policy setting that lend credence to the research designs employed are the following:

- Projects that received funding commitments under HHH were subject to a remarkably consistent set of rules and other constraints. These include

  - a common geography (one large metro area), which reduces the concern that geographic differences may introduce a correlation between PLA use and other area-specific cost drivers

  - a common set of regulatory and other institutional constraints (including dealing with the same agencies and funding characteristics), which eliminates many potential unobservable characteristics of projects potentially correlated with the use of a PLA that may affect costs

  - a common time period (where time is measured by the date of application for LIHTC funding, covering 3.5 years, from June 2017 to December 2020), which eliminates potentially large shifts in, for example, building technologies, labor supply, and other factors that may confound comparisons.[13]

- In addition to the HHH-funded projects, I also incorporate a smaller sample ($n = 34$) of comparable non–HHH-funded supportive housing projects. These projects share the same geographic and regulatory constraints discussed above and occurred in a time period that fully overlaps with the period of HHH-funded projects and only modestly predates it (May 2015 to January 2021).

- Because of the size of the HHH funding initiative, the sample size used in the analysis is large for the condensed geography and time period, allowing for the modeling of construction costs using a relatively rich set of controls while maintaining reasonable statistical power.

The other unique factor related to the validity of the analyses presented in this report is the nature of the threshold that triggers a project falling under the HHH PLA—whether it comprises 65 or more housing units. Typical PLAs associated with the construction of a given project or set

---

[13] As discussed below, I also control for potential shifts in costs over time in a highly flexible, nonparametric fashion that was not used in the past literature reviewed in Chapter Two.

of projects fully specify what is to be built (e.g., a new elementary school) in advance, but they do not specify project costs, which are generally realized through a competitive bidding process. In this setting, the PLA may affect costs, but the nature of the project, in terms of size, location, and most important features, will be unaffected. In the HHH case, project criteria (e.g., project size, location, construction type, target population) are flexible and chosen by developers applying for a funding commitment from HHH. Importantly, one critical project specification chosen by developers, the number of housing units, determines whether or not the PLA applies to the project. Thus, the HHH PLA has the capability to affect not only the cost of building a project but what type of project will ultimately be built, a unique and novel outcome relating to the existing literature on how PLAs affect competitiveness.

In order to present some initial descriptive evidence on the extent to which the PLA did influence what was built and the costs of building, it is first necessary to describe the data more formally.

The Data Used in This Study

The data used in this study represent 132 housing projects (98 HHH-funded projects and 34 non–HHH-funded projects).[14] These data were assembled from multiple sources:

- Initial data—project name, developer information, number and share of units by size, estimated dates of construction start and completion, and basic cost data for both the HHH and non-HHH projects—were collected from spreadsheets provided through the city of Los Angeles's HHH data portal.
- For the HHH projects, additional initial estimated cost data are taken from Galperin (2019).
- Data on intermediate costs and dates, data on the presence of elevators and parking structures, and updated data on number and share of units by size estimated at the time of application for LIHTC funding are taken from a combination of staff reports from the California Tax Credit Allocation Committee (TCAC) for projects applying for 9 percent LIHTC credits and the California Debt Limit Allocation Committee (CDLAC) for projects applying for 4 percent LIHTC credits. For some projects that recently filed applications to these agencies (where staff reports were not yet available), I used data on the joint TCAC/CDLAC application filed by developers.
- Data on parcel size were taken from the website of the Los Angeles County Office of the Assessor.

---

[14] The total data available from the city of Los Angeles comprised 111 HHH projects and 34 non-HHH projects. Of the 111 HHH projects, 13 were non-new construction (either partially or completely rehabilitation of existing housing). These projects were dropped, so the resulting full analysis sample of 132 (98 HHH and 34 non-HHH) projects was used for the estimation of shares of project size in Chapter Four. For the analyses of construction cost in Chapter Five, only a sample of projects for which there was complete data on costs and all control variables was used. This resulted in a subsample of the full analytic sample of 69 HHH-funded projects and 28 non–HHH-funded projects.

- For a subset of projects that have entered the construction phase or are near this milestone ($n = 53$), data on the contract amount (i.e., total construction costs), updated data on the number of units, dates of estimated construction start and completion, and whether or not the project is subject to the PLA were taken from the California Department of Industrial Relations (DIR) website.
- In some cases, additional information on number of stories or other project features were augmented with data from developer websites, architectural firm websites, or public websites focused on the built environment in Los Angeles, such as Urbanize LA.
- Additional data on unit sizes and related details were collected from the Los Angeles Planning Department website.

The resulting data from these sources represents the most complete single dataset on the pipeline of supportive housing projects in Los Angeles that I am aware of. The primary limitation of these data is that the cost component used in this report reflects estimated costs rather than costs after completion. The most ideal data with which to conduct these analyses would be cost-certification data submitted to TCAC/CDLAC after projects are completed and placed in service. This is infeasible since the vast majority of HHH projects are incomplete, with many still as far as two years away from completion, according to estimated dates provided by their developers. However, the estimates used in the main analyses appear likely to reflect ultimate project costs well.

In Appendix A, I provide empirical evidence on the validity of these cost estimates as a proxy for realized project costs. There are two important overall implications of this additional analysis. The first is that the estimated costs used in this analysis change very little for projects that have filed updated costs with the state for wage compliance shortly before commencing construction. The second is that changes in cost estimates do not appear correlated with whether a project is subject to the PLA or not, suggesting that any error in estimating a cost effect of the PLA will be related only to random or "classical" type measurement error, which will only affect the precision of the estimates by increasing the size of the standard errors but not the sign or magnitude of the estimate (Wooldridge, 2020).

Evidence on the Comparability of the HHH and Non-HHH Data Samples

The validity of this study hinges, in significant part, on the comparability of the sample of HHH projects and the smaller sample of non-HHH projects. As mentioned previously, these projects were subject to a large number of similar factors due to being built in Los Angeles and utilizing LIHTC funding. However, if they were to differ in other unobservable ways related to their distribution of project sizes and associated construction costs, this would potentially bias the evidence presented. Two pieces of empirical evidence, however, suggest this is not likely to be the case.

First, Figure 3.1 shows the geographical distribution of these projects across the city. As can be seen, projects from both samples are spread across the city, suggesting that it is unlikely that any substantial bias in these comparisons would arise from HHH-funded projects being systematically located in areas with different cost drivers than non–HHH-funded projects. Second, there is substantial overlap in the actual pool of developers behind HHH and non-HHH projects. Table C.2 shows that, of the 19 unique developers associated with the non-HHH projects in the data, 15 of them (79 percent) are also developing projects using HHH funding. These same 15 developers represent 30 percent of developers building HHH projects. This overlap suggests that selection among the pool of developers in choosing to pursue projects with or without HHH funds is unlikely to be a major source of bias in comparisons.

Figure 3.1. Location of HHH-Funded and Non–HHH-Funded Projects

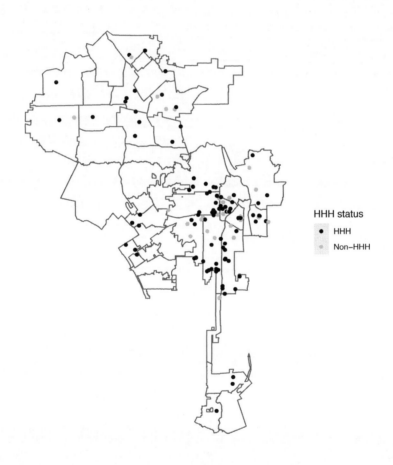

SOURCE: City of Los Angeles, TCAC/CDLAC, and Los Angeles County Office of the Assessor data.
NOTE: Outlines indicate Community Plan Areas in the city of Los Angeles.

Replicability and Research Transparency

In order to promote quality research and to increase knowledge on this important topic, the full dataset and the statistical (Stata) code used in this project (Ward, 2021) have been made available for any researchers interested in replicating the results presented in this study or using the data for further research.

4. Association Between the PLA and Project Size

The goal of Proposition HHH was to induce developers to build thousands of new units of affordable/permanent supportive housing (AH/PSH) by providing a portion of the necessary funding (given realized costs, this has worked out to around 25 percent of total costs per project). The strength of this incentive arises from the fact that AH/PSH developers typically fund projects by building a "capital stack" comprising multiple sources of funding including "soft loans" such as HHH and other sources of state funding (e.g., No Place Like Home, Infill Infrastructure Grant Program, Home Investment Partnerships Program) and LIHTC funding. These soft loans, which tend to have generous terms and take junior positions in terms of repayment relative to other funding sources, must generally be procured first and are often scarce and difficult to secure. Thus, the large funding pool represented by HHH provided a significant head start on building the required capital stack.

This fact, that HHH was a funding program, made it significantly different from the type of programs usually associated with PLAs. A typical program might be, for example, an initiative to fund the construction of 20 new elementary schools, where the plans for the schools are developed in advance and interested contractors bid to build a particular building or buildings. In such a case, the PLA may affect the number of bidders and the amounts of the resulting bids, but it does not change the size or location of the proposed schools. In the case of HHH, a developer was free to submit a proposal for an AH/PSH project that, within certain constraints resulting from both regulations and incentives, could be of a size ranging from 20 PSH units and up serving a range of subpopulations eligible for AH/PSH. But, importantly, any projects of 65 units or more would be covered by the PLA. This difference allowed for the PLA to not only affect the costs of a project but to affect the size of a project.

Graphical evidence that the PLA did indeed affect the size of projects proposed under HHH is provided by Figure 4.1. HHH projects in the data ($n = 98$) are placed into groups of five housing units (e.g., 55–59 units, 60–64 units, 65–69 units) on the x-axis. The y-axis plots the number of projects in each group. Focusing on the two groups nearest the threshold, there are 22 projects in the data with between 60 and 64 housing units while there is one project with between 65 and 69 housing units. In the 60- to 64-unit group, 15 of these 22 projects have *exactly* 64 units. Looking further along the x-axis at the numbers of projects above 65 units that are increasingly further away from the PLA threshold (70 to 94 units), similarly very low numbers of projects are observed. In fact, the number of proposed projects with between 60 and 64 units is equal to the total number of projects with between 65 and 101 units.

Figure 4.1. Frequency Distribution of Project Sizes by Housing Unit for HHH-Funded Projects

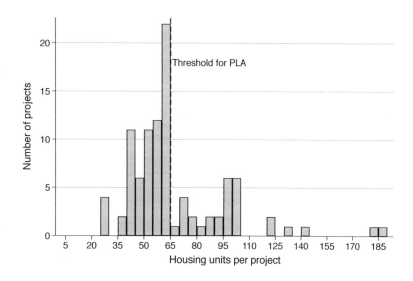

SOURCE: Author calculations using city of Los Angeles, TCAC, and CDLAC data.
NOTE: Data (*n* = 98) are placed into groups of five housing units, with the lower bound of each group denoted on the x-axis (e.g., 50 is 50–54 units).

But how unusual is this distribution? It is possible that there were other factors affecting project size (such as construction type) that also incentivize the proposal of projects of just under 65 units. To provide evidence on the plausibility of such a scenario, we can consider how the distribution of HHH-funded projects by size compares to the distribution of the sample of non–HHH-funded AH/PSH projects built in Los Angeles over approximately the same period. These projects were not subject to the HHH PLA requirement but were generally subject to all other significant factors influencing project scope including land costs, building costs, and state and local regulations.

Figure 4.2 shows the distributions of projects according to the share of the entire sample, using larger 15-unit groupings (with a single group for projects of 110 units or more), for both HHH-funded and non–HHH-funded projects side by side.[15] Note that the share of HHH projects with 34 or fewer units is less than half the size of the same share of non–HHH-funded projects. This is likely driven by the fact that the HHH funding incentivized larger projects.[16]

---

[15] Given the relatively small size of the non–HHH-funded project sample, this aggregation was necessary to have sufficient representation of projects in each group. The size of these groupings was chosen prior to considering the results by splitting at the PLA threshold then using 15-unit groupings. However, alternate groupings do not change the qualitative conclusions from this exercise.

[16] HHH regulations specify a baseline interest rate of 4 percent and incentivized the development of larger projects through progressively larger interest-rate reductions (e.g., a 0.5-percent reduction for developments with 15 to 20 PSH units up to a 3-percent reduction for developments with more than 61 PSH units). Note that this incentive reaches a maximum just below the PLA threshold (City of Los Angeles, 2020).

But the share of projects with 35 to 49 units is approximately identical and the shares of the largest groups (95 units and up) are also relatively comparable. No difference in shares is as large as the difference in the share of projects in the 50- to 64-unit group, which comprises less than 10 percent of the non-HHH projects but more than 45 percent of the HHH projects. This gap is reversed for the shares with 65 to 79 units—where the non-HHH share is approximately twice as large—and 80 to 94 units—where the non-HHH share is around three times as large as the HHH share.

Figure 4.2. Distribution of Project-Size Shares by Funding Source

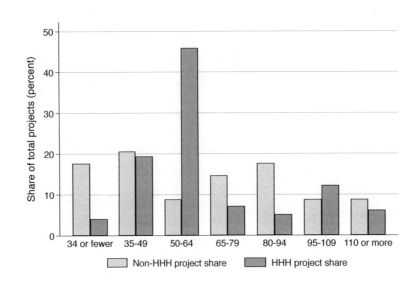

SOURCE: Author calculations using city of Los Angeles, TCAC, and CDLAC data.
NOTE: Data (*n* = 127) are placed into groups of 15 housing units, with the exception of projects with 110 or more units, which are placed into a single group.

This evidence suggests that the disproportionate number of HHH-funded projects proposed to be between 60 and 64 units is directly related to a desire by developers to avoid the PLA. In Chapter Six, I develop a simulation exercise in order to estimate how much the agreement may have affected the overall number of units in the HHH pipeline. In Chapter Seven, I consider why developers may have responded in such a manner.

5. Effect of the PLA on Construction Costs

Existing research on how PLAs affect project costs has typically been confounded by a lack of comparability between projects that did and did not fall under a PLA. Projects covered by these agreements tended to be much larger than those not covered by a PLA and were also often built in major metro areas, while the comparison group of non-PLA projects were often smaller and built in more rural areas, leading to potentially substantial differences in building type, regulation, labor cost, and other important factors that can confound estimation of the effect of PLAs on cost. The current study presented in this report improves significantly on past work by employing a setting where both of these factors are addressed. This study compares AH/SH projects built in Los Angeles over a relatively short time frame that are subject to comparable constraints and costs aside from the presence of one funding source, HHH, that was associated with a PLA. Additionally, since the PLA only affected larger HHH-funded projects, another dimension of cost comparison is feasible: cost differences across smaller and larger HHH projects may be compared with cost differences across smaller and larger non–HHH-funded projects.

I begin by motivating this modeling exercise using descriptive data comparisons to develop intuition about the approach described above. I then introduce and estimate a regression model that puts this approach into a rigorous, plausibly causal estimation framework. I keep the discussion of the model in this text relatively general, but Appendix B contains a more formal description of the model and variables used as well as multiple sensitivity analyses of the results.

A Descriptive Analysis of Costs by Project Size

A key economic concept in the relationship between output and costs is that of *economies of scale*. The idea of economies of scale is that, for many productive endeavors, the cost of producing an additional unit of output decreases as more output is produced. In building multiunit housing, one reason for this phenomenon is that there are considerable fixed costs to building anything at all (e.g., land costs, design and engineering fees, permitting fees, equipment rentals, labor costs) so that once a housing project is to be built, doubling the number of units is unlikely to double the costs because these myriad fixed costs are spread over more units without necessarily increasing in a proportional fashion.

Basic cost tabulations showing how project size is associated with costs for HHH-funded and non–HHH-funded projects can help to illustrate this concept. Panel A of Figure 5.1 presents unit costs according to four groupings of project size for the samples of non–HHH-funded projects and HHH-funded projects. The cost trend across increasing project sizes for the non–HHH-funded projects neatly displays the economies of scale concept. As the average number of units

23

increases, the average per unit total cost declines in a nearly linear fashion from approximately $511,000 to $408,000.

Figure 5.1. Per Unit Costs and Cost Differences by Project Size

Panel A. Per Unit Costs for HHH-Funded and Non–HHH-Funded Projects

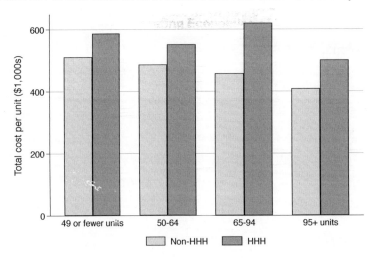

Panel B. Per Unit Cost Differences Between HHH-Funded and Non–HHH-Funded Projects

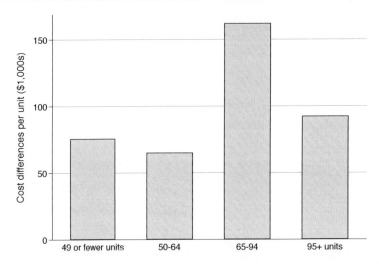

SOURCE: Author calculations from city of Los Angeles, TCAC, and CDLAC data.

By contrast, for the HHH-funded projects, it is notable that these projects are, in each case, more expensive than the similar non-HHH projects (for projects that fall below the PLA threshold, HHH projects average $70,368 more per unit). But note, also, the lack of steadily declining costs among the HHH projects. In fact, the per unit cost of projects with between

65 and 94 units—projects covered by the PLA—is higher than the per unit cost of projects with fewer than 50 units.[17]

To further illustrate this relationship, panel B in Figure 5.1 plots the positive cost differences between HHH-funded and non–HHH-funded projects for each grouping. This approach shows that the cost gap for projects with 65 to 94 units is more than twice as large as the gap between either of the two groups of smaller projects. Even in the group of the largest projects, where economies of scale were most apparent for the non–HHH-funded projects, the cost gap is more than 1.5 times the size of the cost gap between the group of projects with 50 to 64 units of housing.

Illustrating the Research Design Used to Estimate the PLA Cost Effect

Figure 5.2 provides a graphical intuition for the notion of economies of scale and how the PLA may interact with this factor that is analogous to the regression model used to generate the estimates in this study. Panel A plots the construction costs of each HHH-funded project on the y-axis according to the number of units (plotted on the x-axis).[18] A single line is fitted to the data so that the slope of this line represents a linear estimate of economies of scale in construction costs achieved by increasing the number of units.[19] This estimate indicates that each one-unit increase is associated with an average decrease in costs of $232, so that going from either 30 to 60 units or going from 70 to 100 units is associated with a per unit construction cost decrease of $6,960.

By construction, the fitting of a single line to the data does not allow for the estimation of potential differences in construction costs associated with the PLA threshold. It is possible, though, to fit a model with two separate lines, where one line is fit through the data points below the PLA threshold to provide an estimate of the slope of these per unit construction costs, and a second line is fit through the data points above the PLA threshold, providing a separate estimate of the cost slope for these projects. Where these two lines meet (at 65 units), the vertical discontinuity between them represents an estimate of the change in costs associated with building under the PLA versus not doing so. Panel B of Figure 5.2 visualizes these results. The estimated rate at which construction costs for projects of 64 or fewer units declines (the slope of the line on the left side of

---

[17] Other potentially discontinuous changes are possible in the size/construction costs relationship, such as switching to a different construction type (e.g., from wood to steel construction). However, any such pattern should appear in both cost distributions rather than just the distribution of HHH-funded projects. In the regression-based analysis, this issue is at least partially controlled for with data on the number of stories in each project (as described below and further in Appendix B).

[18] To increase the resolution of these figures, I drop the four largest projects, which have between 134 and 187 units, from these plots. For completeness, plots including all units are presented in Appendix C.

[19] This line is the best fit through the data, where best fit is derived from a bivariate ordinary least squares regression model that fits a line by minimizing the sum of squared errors from the data points to the regression line.

the threshold) is twice the magnitude of the rate of decline in panel A ($574 per additional unit versus $232) and the estimated rate at which per unit construction costs decline for projects above 65 units is nearly seven times as large ($1,560 per unit versus $232). The estimated construction cost discontinuity at 65 units is nearly $70,000.

**Figure 5.2. Estimating the Effect of the PLA on Construction Costs
While Incorporating Economies of Scale**

Panel A. Single Slope Cost Estimate and No Allowance for a Discontinuity at 65 Units

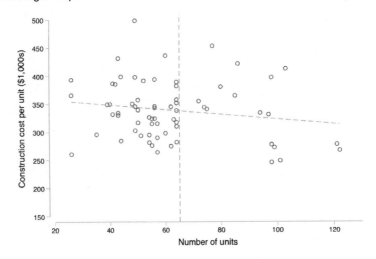

Panel B. Discrete Slope Estimates Above and Below 65 Units

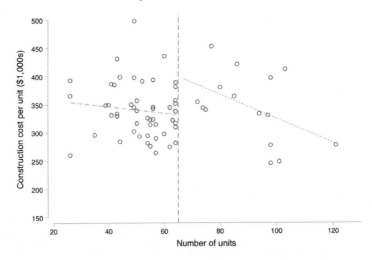

SOURCE: Author calculations from city of Los Angeles, TCAC, and CDLAC data.

It is critical to note, however, that this difference does not control for potentially important factors related to project size that may influence cost and is, thus, not a plausible causal estimate of the effect of the PLA on construction costs. The purpose of this graphical exercise is simply to

26

provide the intuition behind the regression modeling approach used. The model used to generate the actual cost estimates in this report—described below and in more detail in Appendix B—simultaneously uses the discontinuity-based approach at the 65-unit threshold seen in panel B of Figure 5.2 for both HHH projects *and* for the additional sample of non-HHH projects and generates a plausibly causal estimate of the effect of the PLA on construction costs as the difference in these two cost discontinuities (after controlling for the relationship between many potentially important project-specific characteristics and costs). In other words, the model allows for there to also be a discontinuity at this same threshold in the sample of non-HHH projects and then subtracts any such cost discontinuity (that cannot be caused by the HHH PLA) from the estimated cost discontinuity among HHH-funded projects.

Estimating the Effect of the HHH PLA on Costs

The model used in this study to estimate the cost effects of the PLA is closely related to a "difference in regression discontinuities" (DRD) research design that marries the concept of a difference-in-differences (DD) approach (Card and Krueger, 1994) with a regression discontinuity (RD) research design (Thistlethwaite and Campbell, 1960).[20] However, it is important to note that this setting does not meet one important assumption that underlies the RD research design's claim to provide a causal estimate—that units subjected to the policy of interest do not manipulate their position around the relevant policy threshold (McCrary, 2008). As seen in Chapter Four, there is clear evidence that developers chose project sizes in a way that was responsive to the PLA threshold.

For this reason, I focus on an alternate interpretation of the model used to estimate the PLA cost effect that does not hinge on this particular identifying assumption. The model may also be interpreted as a cross-sectional DD model that controls for economies of scale through the inclusion of slope coefficients for the number of units on each side of the PLA threshold. In this case, the key assumption required for validity is that the distribution of project sizes among the HHH sample in the absence of the PLA is well approximated by the distribution of project sizes in the non-HHH sample. Appendix B provides a formal exposition and discussion of this model and the relative merit of these two interpretations.

As mentioned above, to accurately estimate a plausible causal effect of the HHH PLA on construction costs, it is also necessary to control for a variety of potentially important characteristics of these projects that may influence cost and that may be correlated with the size of projects and, thus, with the PLA threshold. The controls included in the regression model (aside from the main "ingredients" discussed in the graphical example in panel B of Figure 5.2 above) are:

---

[20] Recent examples of this approach include Lemieux and Milligan (2008) and Cellini, Ferreira, and Rothstein (2010).

- the number of stories
- whether a project is subject to a requirement to pay prevailing wage or commercial prevailing wage[21]
- the shares of total housing units in each project that are, respectively, studios, one-bedroom units, two bedrooms, and three or more bedrooms
- the share of units that are permanent supportive housing
- the presence of elevators that serve 95 percent or more of the units
- the presence of either an underground or freestanding parking structure
- whether a project's location qualifies it as a transit-oriented development
- variables for the target populations of the included PSH units (individuals, families, and special populations, including survivors of domestic abuse or sexual trafficking, persons with mental illness, seniors, transition-aged youth, HIV-positive individuals, or veterans).

More detail about each of these control variables and the motivation for their inclusion is provided in Appendix B.

In addition to these controls, it is necessary to control for growth in construction costs related to increases in the price of materials and labor. In past research, this has generally been done by adjusting costs of projects (often spanning ten or more years) according to an index of construction industry prices. Because of the compact geography and time period involved in this study, I employ a simpler, nonparametric approach that does not depend on the accuracy of an index of input prices. Rather, I incorporate a series of year fixed effects (indicator variables for each year a project submitted a funding application to TCAC/CDLAC). This approach estimates a common, year-specific, average cost effect for all applications submitted in each calendar year allowing for not only variation in materials prices and wages, but for otherwise unobservable factors that may affect construction costs (e.g., an area labor shortfall relative to demand that leads contractors to price in more overtime pay).

Estimated Cost Effects of the HHH PLA

Table 5.1 presents estimates of the effect of the HHH PLA on construction costs. The results are presented using two approaches to measuring cost differences. Panel A presents results using construction costs in $1,000s as the outcome. Panel B presents results using the natural log of construction costs in $1,000s as the outcome. This approach gives coefficients that can be

---

[21] All HHH-funded projects are required to pay workers prevailing wages and must submit compliance documents to this effect to the California DIR. A subset of HHH projects with certain characteristics must pay a higher wage scale (commercial prevailing wages). For more detail on these controls and how projects are identified with respect to them, see Appendix B.

interpreted in terms of an approximate percent change in construction costs associated with the HHH PLA.[22]

For each panel, I present the results of three specifications of the model. A brief explanation of each of these follows:

- *Specification 1:* This model uses all projects for which complete data on the included control variables could be collected ($n = 97$).
- *Specification 2:* This specification excludes projects below the 5th percentile of project size and above the 95th percentile of project size from the sample so that these outliers in terms of project size do not influence estimates of the cost slopes and, thus, the discontinuity in cost slopes at the PLA threshold, which is the estimate of interest.
- *Specification 3:* This specification addresses the likelihood that project sizes were chosen in response to the PLA threshold and that this sorting around the 65-unit threshold may have been correlated with potentially unobservable differences in a developer's ability to manage construction costs by excluding a portion of the sample from the analysis. I implement this strategy by excluding projects just below the PLA threshold (64 units) and projects that extend a moderate distance (in housing units) above the PLA threshold (to 74 units). This nonsymmetric exclusion removes a total of nine 64-unit projects and three 65- to 74-unit projects from the data. (Alternate approaches to excluding these data are discussed and estimated below.)

---

[22] Coefficients from logged dependent variable models are often interpreted as a percent change by simply multiplying the coefficient by 100, though the more formal percent-change interpretation is given by the following conversion for a given coefficient, β: $(e^{\beta} - 1) \times 100$.

Table 5.1. Estimates of Effect of PLA on Construction Costs

	(1) Full Sample	(2) Exclude Outlier Projects by Size	(3) Also Exclude 64- to 74-Unit Projects
Panel A. Outcome Is Construction Cost in $1,000s			
HHH PLA	44.588[+]	43.344[*]	46.843[*]
	(22.797)	(20.914)	(22.572)
	[0.054]	[0.042]	[0.043]
Adjusted R^2	0.636	0.735	0.759
Panel B. Outcome Is Natural Log of Construction Cost in $1,000s			
HHH PLA	0.195[**]	0.206[**]	0.217[**]
	(0.074)	(0.063)	(0.067)
	[0.010]	[0.002]	[0.002]
Adjusted R^2	0.673	0.792	0.814
Observations	97	86	74

SOURCE: Author calculations from city of Los Angeles, TCAC, CDLAC, and other data sources as described in text.
NOTES: In specification 2, outlier projects are those below the 5th percentile or above the 95th percentile of project size. Standard errors are in parentheses and p-values are in square brackets.
[+] $p < 0.10$, [*] $p < 0.05$, [**] $p < 0.01$.

Considering the series of estimates using these three specifications in panel A, we see that excluding outlier projects by size (going from specification 1 to 2) has little effect on the magnitude of the estimate, but it increases the precision of the estimate as well as the overall explanatory power of the model (measured using adjusted R^2) from 0.64 to 0.74.[23] Specification 3, which excludes projects from 64 to 74 units under the assumption that these projects near the PLA threshold were more likely to have their size influenced by the 65-unit threshold, increases the magnitude of the estimated PLA effect slightly (by around $3,500) while also slightly increasing the overall explanatory power of the model. This modest increase (8 percent) in the size of the estimate suggests that any bias from including these observations is likely downward in magnitude, but that such bias appears to not be a major concern.

The sensitivity of the estimate to the particular data exclusion chosen is explored further in Table B.6, which reproduces specification 3 and compares this result to excluding 64 to 80 units (this drops nine projects on either side of the threshold) or excluding 60 to 90 units (dropping 15 units on either side of the threshold). These alternate results suggest that there is some

---

[23] Adjusted R^2 is a measure of the share of the total variation in the outcome explained by the model after including a penalty for adding additional variables (so that an additional variable can lower the overall explanatory power of the model if it does not contribute meaningfully to improving the model's explanatory power).

moderate sensitivity to how these projects are excluded, but that this sensitivity is all in the direction of increasing the size of the estimate relative to specifications 1 and 2.

Turning to panel B in Table 5.1, which uses the natural log of construction costs as the dependent variable, the results are strongly confluent with the results using the outcome variable in dollar terms, but the estimates are generally more precise, with all results statistically significant at the 99-percent confidence level. Additionally, specification 3 explains more than 80 percent of the total variation in the dependent variable. Note that applying the average estimated effect across these three specifications of around 21 percent to the mean of construction costs for 50- to 64-unit projects ($328,789, the lowest average per unit construction cost among the sample of HHH projects) implies that the PLA increased construction costs by more than $69,000 per unit.

With respect to considering the performance of this model more broadly, all of these specifications explain a substantial portion of the total variation in construction costs despite a total sample size of less than 100 and the inclusion of more than a dozen control variables. Additionally, the coefficients for the HHH PLA term are all highly precise, with all three estimates statistically significant at or near the 95-percent confidence level. This is also the case for nearly every sensitivity test presented in Appendix B, including those that discard up to 30 percent or more of the sample data, suggesting a persistent, robust statistical relationship.[24]

Appendix B contains a wealth of additional material relating to these estimates. Tables B.1 and B.2 present results for each specification with coefficients and standard errors for all included controls. Results and discussion of the exclusion of two large projects that were proposed before the adoption of the PLA (and thus were not subject to the PLA requirement, despite being over 65 units) and the effect of including these projects on the estimated cost effects are presented, respectively, in Table B.4 and Appendix A. Table B.5 presents results for two alternate versions of the preferred specification that use either a single quadratic term to estimate the per unit cost slope or two separate quadratic terms for either side of the PLA threshold. Table B.6 presents specifications that exclude differing amounts of data around the PLA threshold. None of these sensitivity tests has a qualitatively meaningful effect on the estimated effect of the PLA, either in terms of magnitude or precision. Finally, Table B.7 provides descriptive statistics for all of the variables included in the model grouped by HHH status.

---

[24] Regarding statistical significance as a marker of whether to accept or reject a finding, while more than a century of habit has led to an unscientific focus on using these arbitrary thresholds in a pass/fail manner with respect to the results of a statistical model, well-deserved scrutiny of this practice is growing. I reference this common threshold to facilitate comparison with past research but note that, in nearly every case, the precision of these results provides strong evidence in favor of rejecting a null hypothesis that the effects of the PLA on construction costs are at or even near zero. For a more thorough discussion of this issue, including recent proposals in the field of applied statistics to abandon the paradigm of statistical significance entirely, see, for example, McShane et al. (2019).

Interpreting Model Results

Because of the lack of a strong conceptual motivation to use logged costs (as would be the case if, for example, project costs displayed an exponential relationship with project size) and in order to take a relatively conservative approach to interpreting these various estimates, I focus discussion of these results on the estimate from specification 2 using dollars in $1,000s as the outcome, rather than either specification 3, with the additional data restriction, or the uniformly larger (and more precise) estimates using the logged dependent variable.[25] This estimate indicates that the HHH PLA added approximately $43,000 per housing unit to projects covered by the agreement.

Figure 5.3 puts this cost increase in percentage terms using two different sets of comparison projects. The left-hand bar shows an estimate of the increased PLA cost in percent terms relative

Figure 5.3. Interpreting HHH PLA Cost Estimates by Comparison with Larger Non-HHH Projects or Smaller HHH Projects

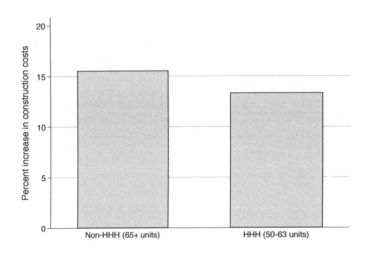

SOURCE: Author calculations from city of Los Angeles, TCAC, and CDLAC data. See text for detailed description of calculations.

to the average construction cost of non-HHH projects above the PLA threshold ($253,877 per unit), since these projects will reflect construction type and other factors common among larger projects.[26] This approach indicates that the HHH PLA increased construction costs by 15.5 percent.

---

[25] The "log-linear" specification is quite standard in the literature. The data exclusions in specification 3 are justified on theoretical grounds, but the empirical evidence for assuming that these projects were chosen endogenously is lacking, as is a good guide for where exactly to limit this data restriction in terms of size. The exclusion of outlier data points in terms of size is also quite common in many empirical studies, mimics the "local linear regression" approach taken in most RD studies, and also happens to be the smallest of the several estimates considered.

[26] This per unit cost is also adjusted upward by the average positive difference in construction costs among all HHH projects relative to the sample of non-HHH projects since the model results (see Appendix B) indicate there was an additional average cost difference across all HHH-funded projects that was unexplained by the presence of the PLA.

Another approach shown on the right-hand side of Figure 5.3 gives an estimate of the increased PLA cost in percent terms relative to the construction costs of HHH-funded projects just below the PLA threshold (I take this measure from projects of between 50 and 63 housing units, which had an average per unit construction cost of $324,585). These projects are the largest HHH-funded projects not subject to the PLA and reflect other cost factors that may be common to all HHH-funded projects around the threshold. Here the estimated PLA cost effect is 13.4 percent. Taking the simple average of these two estimates suggests that the PLA increased per unit construction costs of affected projects by 14.5 percent.

How plausible is it that these estimates accurately reflect the causal effect of the PLA on costs? The improvements in the research design relative to past studies suggests that this estimate is more likely to reflect the true construction cost effect of a PLA. But, taking the validity of the past studies considered earlier at face value, a roughly 16-percent increase in costs is on the lower end of prior results, which have generally ranged between 15 and 20 percent. The highest estimates from my analysis (using the logged construction cost approach) are in line with the higher results (around 21 percent) found in this body of past research.

One useful falsification test for these results involves using soft costs instead of construction costs as an outcome. While soft costs are indirectly related to construction costs through the inclusion of construction contingency funding and the developer fee, the link between these components and the PLA should be weak. For other components of this category, such as architectural and engineering fees, there may be a link between costs and project size, but it should not jump in a discrete fashion at 65 units, and discrete changes in such fees related to, for example, changes in construction type should also be addressed by the model's controls. Components such as carrying costs of financing should not be meaningfully affected by whether or not a project contains specifically more or fewer than 65 units. Still other components, such as costs related to service provision, may demonstrate economies of scale that could lead to costs declining as the number of housing units increases. For these reasons, the effect of the PLA on soft costs should not be similar to the main results using construction costs.

These results are presented in Table B.3, and they are supportive of the validity of the main results. Unlike the results for construction costs, which are relatively smaller and imprecise in specification 1 and increase in magnitude and precision from there, these results are relatively larger and more precise in specification 1 but become smaller and less precise by specification 3 even as the overall explanatory power of the model increases, suggesting that correlations between soft costs and the PLA are spurious and not robust to a well-specified model.

This adjustment has the effect of reducing the estimated difference in this comparison, which I view as a conservative approach.

Limitations of the Cost Analysis

While this analysis improves on most past research in both the setting and the approach employed, there remains the possibility that I do not fully capture all relevant factors that may contribute to both the choice of project size and the cost of projects, including the availability and terms of financing, differences in costs associated with site preparation, evolving regulatory regimes, and other factors. Additionally, I do not explicitly control for construction type or for the amount of commercial space, the amount of nonresidential space for, as examples, medical or mental health services, and other similar factors that may influence project size and costs. However, I have attempted to provide all feasible and relevant measures related to such potential sources of bias and have attempted to discuss the likelihood that these are of concern in a transparent manner. In considering what threats to the validity of these estimates may remain, I note that, given the idiosyncratic nature of the PLA threshold, factors likely to significantly bias the estimated PLA effect would have to be explicitly correlated with project size around the PLA threshold or would have to be correlated meaningfully with construction costs above or below this threshold.

6. Synthesizing Cost and Output Effects

The findings in Chapters Four and Five are each important and relevant to policy, but to meaningfully consider the costs of the HHH PLA in a more holistic sense requires a plausible counterfactual outcome that incorporates the agreement's effects on both the number of housing units produced and the cost of producing them. In this chapter, I conduct a simulation exercise that serves this purpose. Broadly, this involves predicting a cost for each project in the data after "turning off" the PLA component of the regression model. Then, using alternative assumptions over the shares of projects according to the number of housing units in each one, a sample of HHH projects satisfying these assumptions is drawn at random. The results of this exercise indicate that in a world with Proposition HHH but no HHH PLA, approximately 800 more units of housing could have been produced with the same level of expenditures. This is equivalent to 11 percent of the total number of housing units currently in the HHH pipeline.

This simulation exercise proceeds in the following manner. First, I generate a counterfactual per unit construction cost for each project in the HHH sample by using the model results from the preferred specification above (specification 2 expressed in dollar terms), setting the indicator variable for the PLA to zero (in essence "turning off" the PLA), and then predicting a counterfactual construction cost for each HHH project in the data under the assumption that there was no PLA. This counterfactual construction cost is then added to the actual land and soft costs to create a counterfactual total per unit cost.[27]

Second, using this sample with a counterfactual cost generated for each project, the distribution of projects is divided up into size groupings (identical to those used in Figure 4.2) and a new sample of HHH projects equal to the number of projects in each share in the actual analysis sample is drawn at random using two different assumptions about how the size of projects would have been distributed in the absence of the PLA. In both cases, I assume that the PLA did not affect project size for projects of fewer than 50 units or projects of more than 94 units; projects for these shares are still drawn at random, but the shares are fixed at the share observed in the HHH analysis sample (these fixed shares make up around 45 percent of the total sample).

---

[27] For this exercise, I add back in the two excluded large, non-PLA projects discussed elsewhere in this report and use their predicted construction costs without adjusting PLA status. Due to their sizes, they are only used in draws for the static shares of projects of between 95 and 109 units and 110+ units. Their inclusion is motivated by increasing the pool of projects of different sizes available for drawing simulated subsamples of these larger-sized shares.

Table 6.1. Observed and Alternative Distributions of HHH Project Shares by Size in Percent Terms

Units	15–34	35–49	50–64	65–79	80–94	95–109	110+
Observed (HHH sample)	4.1	19.4	45.9	7.1	5.1	12.2	6.1
Scenario 1 (uniform shares)	4.1	19.4	19.4	19.4	19.4	12.2	6.1
Scenario 2 (non-HHH shares)	4.1	19.4	12.5	20.8	24.9	12.2	6.1

SOURCE: Author calculations from city of Los Angeles, TCAC, and CDLAC data.
NOTE: Each value reflects the share of projects in either the observed or counterfactual distributions of projects by grouped sizes expressed in percent terms. Size groupings shaded in light gray are the shares that are altered in the simulations.

For projects between 50 and 94 units, I assume in scenario 1 that the total share of projects in the actual HHH sample, which are disproportionately in the 50- to 64-unit group, would instead be uniformly distributed across the three affected size groupings (50–64 units, 65–79 units, and 80–94 units) in the absence of the PLA. In scenario 2, I assume that these projects would instead be distributed according to the relative shares of each project size observed in the actual non-HHH project sample (which has increasing shares of projects moving across these three groupings). Table 6.1 presents, first, the observed distribution of projects by share in each size group, then the counterfactual shares assigned to each group according to the two alternative scenarios described. The groups in which shares are reallocated are shaded in light gray.

Using these three sets of project size shares, alternate samples of projects are created using a basic bootstrap method that randomly draws (with replacement) observations from each size grouping until a simulated group of projects of the appropriate size (according to the shares in Table 6.1) is created for each of these grouped size categories. For each new simulated sample of projects, the sum of housing units and the means of construction costs and total per unit costs are calculated and saved. This process is repeated 1,000 times and then the mean values of these 1,000 descriptive statistics from the simulated project samples are calculated.[28]

This method is used, first, to simulate the actual data (using the "Observed" shares in Table 6.1). The goal of this simulation is to test how well the model performs in regenerating the observed distribution of units and costs. Then it is used with the shares given in scenarios 1 and 2 to generate simulations that simultaneously estimate the total number of housing units and the cost of building them in a counterfactual setting where there was no HHH PLA.

Table 6.2 presents the results of this simulation. The first column shows the total number of units, the per unit construction cost, and the total per unit cost of the projects in the HHH analysis data sample. To the right of these values are three sets of simulated outcomes. The first

---

[28] One limitation of this relatively simple model is that it cannot simultaneously simulate counterfactual numbers of PSH units and total AH units. Under HHH regulations, the share of PSH units must be at least 50 percent, but in practice, the share is a skewed distribution with a mean value of 79 percent and a median value of 93 percent. Thus, modeling these two shares simultaneously would have required more complex assumptions around how the PLA may have influenced the choice over this share, though it was not specifically targeted by the PLA threshold.

simulation replicates the observed data. Each simulated value is within 1 percent of the observed value, suggesting that the model is well calibrated.

In the first row, the two simulation scenarios provide specific estimates of how many more housing units would have been produced under the two assumptions on the redistributed shares of projects outlined in Table 6.1. In the first scenario, where the assumption was that the total share of projects with between 50 and 94 housing units was reallocated uniformly across the 15-housing-unit groupings (50–64 units, 65–79 units, and 80–94 units), the number of housing units built is predicted to increase by 9.1 percent. In the second scenario, which mimics the relative shares observed in the non-HHH sample of projects where there are relatively more large projects within this range, the number of housing units is predicted to increase by 9.7 percent.

Table 6.2. Simulation of HHH Housing Costs and Output in the Absence of the PLA

	Observed Values of Analysis Sample	Simulation of Observed Values		Simulation Scenario 1		Simulation Scenario 2	
		Estimate	Percent Difference	Estimate	Percent Difference	Estimate	Percent Difference
Total units	4,796	4,801	0.10	5,233	9.11	5,263	9.74
		[62.99]		[60.06]		[61.95]	
Construction cost per unit	339,328	336,401	−0.86	324,132	−4.48	323,930	−4.54
		[4,666]		[4,531]		[4,514]	
Total cost per unit	559,492	558,718	−0.14	551,016	−1.51	550,978	−1.52
		[8,708]		[7,651]		[7,728]	

SOURCE: Author calculations from city of Los Angeles, TCAC, and CDLAC data.
NOTE: "Percent difference" column is the difference between observed values of the analysis sample and the mean of the distribution of simulated values. Standard deviation of the distribution of each simulated value is in square brackets.

Turning to simulated results for costs, the estimates are approximately identical, both indicating that average construction costs would decline around 4.5 percent and average total per unit costs would decline by approximately 1.5 percent across all units produced. Note that these differences reflect changes in cost averaged across *all* HHH projects in the absence of the PLA, not only projects covered by the agreement.

Since this simulation uses a representative but incomplete sample of projects from the total HHH pipeline, extrapolating these results to the observed current housing output of HHH requires applying these percentage change values to total number of units in the current HHH pipeline. Table 6.3 presents these calculations using data posted by the city of Los Angeles as of March 2021. Row 2 of the first column shows the number of units in the current HHH pipeline, 7,305. In the first column for each of scenarios 1 and 2 ("Housing output and associated savings"), the measure of total housing units in the HHH pipeline is increased by the percentage indicated in Table 6.2 (row 1). Row 1 shows the number of additional predicted housing units represented by this percent change. Row 3 multiplies the average cost savings per unit given in Table 6.2 by the number of units in the current HHH pipeline to show the predicted savings under the lower costs predicted in the absence of the PLA. To further clarify the meaning of these columns, these results represent, first, how many housing units would have been produced if there was no PLA with a 65-unit threshold for developers to respond to and, second, how much money would have been saved because of the lower costs of producing this collection of housing units without the PLA adding costs to larger projects (of which there would be more). Finally, row 4 displays the percent difference between the predicted total number of housing units and the number of housing units in the observed HHH pipeline.

In the second set of columns for each of scenarios 1 and 2 ("Total housing output with savings spent on additional housing units"), the cost savings in row 3 of the "Housing output and associated savings" column is converted into additional housing units using the average total per unit cost for each scenario in Table 6.2. Now, row 1 of each of these columns indicates the costs of the PLA entirely in terms of additional housing units and row 2 indicates the total housing output of HHH, absent the PLA, under each scenario.

Taking a simple average of the results of these two simulation scenarios indicates that 811 more units could have been produced with the level of funding that has been allotted under HHH. This represents an increase of around 11 percent from the observed number of units. Notably, these estimates of the total number of housing units predicted to have been produced in the absence of the PLA (between 8,093 and 8,139) exceed the more conservative estimate for total housing output (8,000) that was proposed during the HHH campaign.

Table 6.3. Imputing Simulated Cost and Unit Changes to Full HHH Pipeline

	Observed HHH Pipeline	Scenario 1		Scenario 2	
		Housing Output and Associated Savings	Total Housing Output with Savings Spent on Additional Housing Units	Housing Output and Associated Savings	Total Housing Output with Savings Spent on Additional Housing Units
Additional housing units		666	788	711	834
Total housing units produced	7,305	7,971	8,093	8,016	8,139
Total cost savings (millions)		$67.6	$0	$68.3	$0
Percent increase in total housing units		9.1	10.8	9.7	11.4

NOTE: Calculations reflect the application of the percent changes in the simulation sample applied to the total number of units in the HHH pipeline, according to March 2021 data from the city of Los Angeles.

7. Discussion and Policy Considerations

Briefly: A Broader View of Proposition HHH

While this report is focused on a component of Proposition HHH that contributed to reducing the housing output associated with the initiative, it is worth pausing for a moment to reflect on a number of important achievements Los Angeles has made toward addressing the crises of homelessness and housing affordability facing Angelenos that are either directly related to HHH or were likely aided by the attention and resources associated with HHH. Among these are a dramatic scaling up of the production of PSH in the city. The current amount of housing in the HHH pipeline, over 7,300 units (and over 5,700 units of PSH), represents a significant achievement. Furthermore, the HHH Housing Innovation Challenge has advanced the development of a number of alternative approaches to producing PSH including modular construction techniques, adaptive reuse, and alternative funding approaches (Fiore et al., 2019). The attention that HHH focused on providing this type of housing also likely led to other actions addressing a number of barriers to building AH/PSH. Examples of such progress include the following developments:

- The explicit use by city council members of "pocket vetoes" as a passive way to block such projects from being situated in their districts was curtailed (Alpert Reyes, 2018).
- Two related bills, a city of Los Angeles PSH ordinance and Assembly Bill 1197, were adopted to exempt PSH and 100 percent AH developments from costly litigation under the California Environmental Quality Act (CEQA), a longtime tool of opponents of a wide variety of urban infill projects (Hernandez, 2018)
- The city planning department raised the threshold for project size triggering time-consuming site plan reviews from 50 to 120 units (200 in certain parts of downtown LA), as well as other entitlement reforms (City of Los Angeles, 2017)
- Six months after the passage of Proposition HHH, Measure H, a countywide ballot initiative that provided funding for the supportive services that are integral to PSH, was also passed (Conrad N. Hilton Foundation, 2019).

These significant achievements are worth bearing in mind as this and other research focuses on using the lessons of HHH to improve the effectiveness and transparency of future large-scale fiscal policies to address the housing affordability and homelessness crises in Los Angeles.

Summary of Study Results

The results of this study indicate that developers responded to the HHH PLA by proposing a dramatically smaller number of projects at or above 65 units. A total of 22 projects awarded HHH funding (around a quarter of the total) were between 60 and 64 units, while only a single project was proposed with between 65 and 69 units. Comparison of these shares with a similar sample of projects not funded through HHH suggests that the HHH PLA was the causal factor behind this difference.

The results also indicate that the HHH PLA was associated with a per unit cost increase of approximately $43,000. This amounts to a 14.5-percent increase in construction costs and an 8-percent increase in overall per unit costs for projects subject to the PLA.

In a simulation exercise, I estimate that in the absence of the PLA, approximately 800 additional units of housing could have been produced with the funding that has been allocated to date. This represents an increase of approximately 11 percent of the total number of housing units currently in the HHH pipeline.

Why Did Developers Avoid the PLA by Proposing Smaller Projects?

It is unclear why developers responded so strongly to the presence of the PLA. The small developer community that builds supportive housing projects in the region is composed primarily of nonprofit, mission-driven developers, so the type of profit motive that might motivate traditional developers of market-rate housing is largely absent in this setting.[29] The relatively small number of general contractors that are building most of the HHH-funded projects are for-profit, though many successfully bid on both PLA and non-PLA projects, and contractors do not ultimately decide project sizes.

Concerns about the PLA adding uncertainty to costs and timelines may have been an important factor. Deeply subsidized affordable housing projects already face considerable uncertainty related to community opposition, assembling the necessary funding, and uncertain timelines for regulatory approvals. A survey of developers building HHH-funded projects found that complexity directly related to participating in HHH was perceived as the most challenging aspect of their project (Duong, 2021). It may be that the PLA was one source of HHH-related uncertainty that was avoidable through the choice of development size.

It may also be the case that developers estimated that cost increases associated with crossing the PLA threshold prevented a larger project from penciling (i.e., from being estimated to be financially viable). Limits on LIHTC funding as well as caps on other funding sources and an inability to "stack" multiple funding sources (i.e., use them together on one project), a feature common to a number of state-run funding programs in California, may significantly constrain the overall maximum costs of a project (California Department of Housing and Community Development, 2020).

---

[29] In the analysis dataset, 84 percent of the developers or developer partnerships consist of one or more nonprofit organizations.

Are PLAs an Effective Way to Foster the Employment of Local and Disadvantaged Angelenos?

As noted earlier, the inclusion of a PLA with Proposition HHH was motivated by the city council wanting to ensure that HHH spending would contribute to the social and occupational mobility of local and disadvantaged city residents. These goals are expressed in the "targeted hiring provisions" included in the PLA requiring that local workers identified by zip code and qualifying conditions (such as veteran status, chronic unemployment, low income, criminal justice involvement, single parenting but also, less intuitively, being a union apprentice with less than 15 percent of the hours required to graduate to journey level) perform a plurality of the work on covered projects. However, unlike other requirements of the PLA around hiring and workforce issues that have a well-defined set of procedures for enforcement and arbitration, the THPs in the HHH PLA (and similar PLAs in Los Angeles and elsewhere) have no explicit enforcement mechanisms associated with them. The language concerning these goals requires unions to "exert their best efforts" and to "encourage" utilization of targeted workers (City of Los Angeles, 2018).

If ensuring a sufficient level of local and targeted hiring is the primary goal of public works PLAs, alternative approaches such as "first source" hiring programs may achieve these goals in a more flexible fashion. These ordinances cover how jobs are advertised, set project-specific levels of local hiring and work performance goals for publicly funded construction projects, and use enforceable requirements including financial incentives and penalties (Cantrell and Jain, 2013). Los Angeles has an existing first source hiring ordinance with financial enforcement mechanisms that applies to a variety of city contracts (City of Los Angeles, 2016). Additionally, the neighboring municipality of Pasadena has a first source ordinance that applies to construction projects (City of Pasadena, 2021). Such programs are not conditioned on union membership—an important fact since unionized workers make up only around 25 percent of the state's total construction workforce—and can, thus, draw on a much larger and more diverse pool of local workers. Table 7.1 provides demographic information for the state's construction workforce from 2019, showing both the share of workers by race/ethnicity and, separately, by gender, as well as the share of each worker subgroup that is unionized.

If, instead, a primary policy goal is to ensure that publicly funded housing is built primarily or exclusively using a unionized workforce, then directly including such language in future ballot initiatives or other proposals, rather than adding it as a post hoc regulation, would contribute to more fully informed decisionmaking among the public on important fiscal matters. Furthermore, being transparent during debates over large-scale housing policies about the trade-offs involved in attaching labor regulations that restrict the construction workforce eligible to build publicly financed housing may help to set realistic expectations and to avoid the kind of erosion of public and policymaker support for Proposition HHH commonly expressed in the media, in the city council, and elsewhere in recent years.

Table 7.1. Demographics and Union Status of California Construction Workers, 2015–2019

	White, Non-Hispanic	Hispanic	Black	All Other	Female (all)
Number of workers	192,488	325,542	13,718	42,260	11,828
Share of total construction workers	33.3	56.3	2.4	8.0	2.9
Union share *within* group	37.6	19.2	32.5	22.9	22.6

SOURCE: Author calculations from 2015–2019 American Community Survey (ACS) 1-year estimate data (Ruggles et al., 2021) and 2015–2019 Current Population Survey (CPS) Basic Monthly survey data (Flood et al., 2020). NOTE: ACS tabulations of number and share of construction workers in each demographic group use person weights. CPS tabulations of the union share within each group use earnings supplement weights. Construction workers are identified in both data sets as those in the "Construction Trades" occupational grouping and the "All Construction" industry grouping (1990 definitions). The union share counts both union members and those covered by a union contract. The total sample sizes are 24,774 for the ACS data and 711 in the CPS data. Despite the small sample size, the CPS data yield demographic shares of construction workers similar to the ACS shares (29.7 versus 33.3 for white, non-Hispanic, 59.4 versus 56.3 for Hispanic, 2.3 versus 2.4 for Black, 8.6 versus 8.0 for All other, and 1.9 versus 2.9 for female), suggesting that the CPS-based union share measures are reasonably accurate. The race/ethnicity classifications used above combine the "race" and "hispan" variables in the following way. For the ACS, respondents who self-identify as "white" and as "non-Hispanic" are classified as "White." Respondents who self-identify as either "white" or "other race" and as ethnically Hispanic are coded as "Hispanic." Respondents who self-identify as "Black" are coded as such, and the "all other" is the residual grouping. For the CPS, the approach is similar except that Hispanic is simply the intersection of those who self-identify as "white" and "Hispanic" (the CPS data are not characterized by a large subset of respondents identifying as ethnically Hispanic and non-white, as is the case in the ACS).

Why Do These Findings Matter?

In the near term, the cost estimates in this report may aid in ongoing decisions regarding the expenditure of any remaining HHH funds. They may also provide some clarity on the extent to which the PLA may justify requested project cost increases. Reports from the HHH Administrative Oversight Committee have cited the PLA as one of multiple factors justifying cost increases for specific projects that range from $16,000 to $125,000 per unit (Cervantes, 2019; Sewill, 2021).

Moreover, these estimates can also serve as a benchmark against which to measure outcomes related to the HHH PLA and any similar PLAs used in initiatives focused on AH/PSH production. Such analysis can help inform debate over the costs and benefits of using PLAs to meet such goals in future programs compared with alternative approaches that have less of an effect on the primary affordable housing production goal. Additionally, these results may provide greater insight into the potential costs and benefits of pending and future housing legislation requiring the use of a "skilled and trained" workforce, a requirement that is similar in many respects to a PLA in terms of both potential costs and potential behavioral responses from AH developers. The addition of "skilled and trained" language to virtually all bills related directly or indirectly to housing production is currently a major focus of labor union representatives in Sacramento (Koseff, 2020; Mai-Duc, 2021).

Areas for Future Research

The issues raised in this report point to a number of areas where evidence would be helpful in the creation of effective housing and labor policy. Topics of high value include

- surveying developers about their knowledge of PLAs and other labor regulations that are currently being used or considered, including PLAs and "skilled and trained" workforce requirements, which could increase our understanding of the developer behavior observed in response to the HHH PLA and inform future policy design that accommodates important cost constraints or other factors that may be triggered by these types of labor policies
- revisiting HHH costs once a sufficient sample of projects are completed and placed into service using cost-certification data from the state of California to see whether estimates used in this report are accurate and whether any gaps in estimated and actual costs close over time as developers in the AH community gain experience completing projects
- collecting and analyzing data on the utilization of union contractors and subcontractors among HHH-funded and non–HHH-funded projects in order to assess whether the PLA was associated with an overall increased level of union labor or whether reductions in the number of larger projects may have actually reduced overall union labor utilization
- generating meaningful evidence on the landscape of hiring and utilization of local and disadvantaged workers in the construction trades on non-PLA job sites and how this compares to job sites governed by a PLA or related regulations currently being considered or adopted (e.g., skilled and trained workforce requirements).

Concluding Remarks

This report contributes timely evidence that may help guide future spending on the production of AH/PSH, as funding for such efforts appears poised to grow dramatically at the state and local level in the near future. Mayor Eric Garcetti has proposed spending $362 million on additional AH/PSH units (Oreskes and Zahniser, 2021). At the state level, the governor's budget proposal is said to fund the provision of housing for 65,000 people, but whether such a goal can be met depends critically on understanding how policy implementation affects costs (Warth, 2021). More generally, understanding the effect of incentives implicit in labor and housing policy and providing evidence necessary for informed debate is critical for maintaining the public support that will be required to solve our homelessness crisis and meet the challenge of making California an affordable and equitable place to live.

Appendix A: Descriptive Statistics and Discussion of Cost Data

Descriptive Statistics on Costs

Table A.1 provides summary statistics for overall per unit costs and the three major categories of project costs. For HHH projects, average HHH funding commitments per unit are also included. I first present overall average costs, then summarize these same costs for subgroups of projects grouped by unit size. The first group is projects with fewer than 50 units, the second is projects from 50 to 64 units (just below the PLA threshold), the third is projects from 65 to 94 units (above the threshold), and the fourth is projects of 95 or more units.[30]

---

[30] Aside from the split at the 65-unit PLA threshold, these groupings are arbitrary, but the results are not qualitatively affected by other groupings. Two such alternate groupings are considered in the replication code that accompanies this report.

Table A.1. Costs of Projects Grouped by Number of Units

	All	Fewer than 50 Units	50 to 64 Units	65 to 94 Units	95 or More Units
Panel A. HHH Projects					
Unit cost	562,337	587,027	551,615	619,769	508,141
	[85,355]	[80,274]	[61,958]	[92,178]	[114,427]
Construction cost	341,342	357,273	328,789	374,033	325,427
	[55,391]	[55,507]	[40,623]	[42.523]	[83,649]
Land cost	51,154	44,814	53,185	66,401	45,292
	[30,703]	[33,492]	[28,046]	[31,502]	[32,000]
Soft cost	169,841	184,940	169,640	179,335	137,422
	[39,701]	[34,031]	[33,426]	[54,450]	[39,301]
HHH cost	152,281	179,999	157,375	134,488	102,991
	[52,306]	[47,878]	[53,781]	[33,982]	[22,948]
Average units	66.3	40.9	57.9	80.4	123.7
	[31.1]	[7.6]	[4.9]	[7.5]	[33.7]
Observations	69	19	31	8	11
Panel B. Non-HHH Projects					
Unit cost	470,838	511,391	486,515	457,677	408,393
	[105,346]	[103,461]	[112,178]	[120,805]	[59,994]
Construction cost	258,722	263,451	268,129	266,338	233,925
	[62,498]	[47,393]	[33,854]	[94,594]	[40,762]
Land cost	55,917	59,415	56,056	49,903	58,455
	[30,876]	[37,747]	[43,262]	[26,891]	[21,987]
Soft cost	156,199	188,526	162,330	141,436	116,014
	[44,397]	[40,692]	[40,518]	[37,282]	[13,940]
Average units	68.2	37.2	55.0	82.9	109.5
	[30.3]	[10.2]	[7.8]	[7.9]	[11.8]
Observations	29	11	3	9	6

SOURCE: Author calculations from city of Los Angeles, TCAC, and CDLAC data.
NOTE: Standard deviations for each measure in square brackets. Two projects with a CA Department of Industrial Relations PLA status that does not follow the 65+ threshold rule are omitted.

Assessing the Validity of Estimated Cost Data

As discussed in Chapter Three, the ideal data source for estimating cost differences associated with the PLA would be cost certifications provided to the state agencies (TCAC and CDLAC) that allocate LIHTC funding after the completion of all the relevant projects. However, there are multiple factors suggesting that the cost estimates used in this study are sufficient to provide an unbiased estimate of the effect of the HHH PLA on construction costs.

Figure A.1. Average Estimated per Unit Construction Costs for HHH Projects over Time

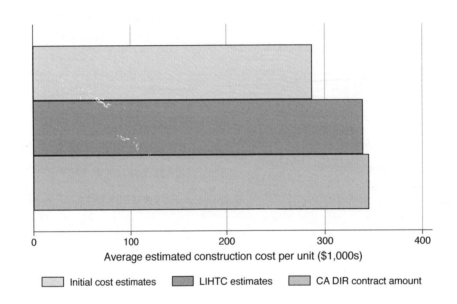

SOURCE: Author calculations from city of Los Angeles, TCAC, CDLAC, and CA DIR data.

Figure A.1 plots the average construction cost estimates submitted by developers for a subset of HHH projects ($n = 53$) that entered or were about to enter the construction phase. Such projects are required to submit data to the California DIR for the purposes of monitoring compliance with state labor regulations during the construction phase.[31] I measure these estimated costs at three different time periods: (1) during the initial proposal to the city to secure an HHH funding commitment, (2) during subsequent proposals to the state agencies that award LIHTC funding (the primary data used in this study), and (3) shortly prior to the start of construction. These correspond to the three cost estimates in Figure A.1.

As can be seen, estimated costs increased significantly from the initial estimates provided by developers to the city to secure an HHH funding commitment to the subsequent estimates

---

[31] The CA DIR cost estimate is taken from the "contract amount" field for registered projects. These data were collected using the Public Search Utility tool (California Department of Industrial Relations, 2010).

provided to the state tax committees that allocate LIHTC funding (an 18.1-percent increase). However, estimated costs only increased another 1.8 percent between the time that cost estimates were provided for obtaining LIHTC funding and when they were provided to the California DIR. The fact that these cost estimates appear to vary little between when data are provided to the state committees allocating LIHTC funding and when these projects are about to begin construction suggests that the detailed LIHTC cost estimates provide a good approximation of eventual project costs.

Figure A.2. Percent Change in Estimated per Unit Construction Costs over Time by PLA Status

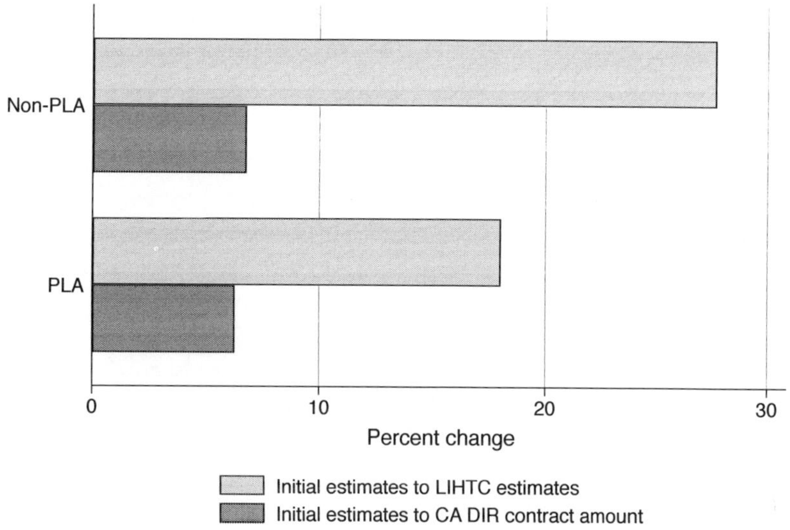

SOURCE: Author calculations from city of Los Angeles, TCAC, CDLAC, and CA DIR data.

Additionally, in terms of inaccurate cost estimates creating bias in the estimated effect of the PLA on construction costs, the only requirement is that any error in these estimates is not directly correlated with the PLA threshold itself. As shown in Figure A.2, this does not appear to be the case. The first set of bars (in light gray) show that from the initial city estimates to LIHTC estimates, construction costs increased by substantially more for non-PLA projects than for PLA projects (28 percent versus 18 percent), but that from the LIHTC estimates to the CA DIR contract amounts, the change was 6.7 percent for non-PLA projects and 6.2 percent for PLA projects (this difference is statistically imprecise in a regression of the difference in costs on PLA status, with a p-value of 0.93).

Finally, there are incentives inherent in the LIHTC funding process that mitigate against developers using highly inaccurate cost estimates. Specifically, TCAC penalizes a project that has final costs that greatly exceed estimated costs; see section 10317(i)(5) of the California Code of Regulations (2020).

47

Effects on Costs of Including Two Large Projects Not Subject to the PLA

In the analysis sample used in this study, there are two projects that have a PLA status that do not follow the threshold rule expressed in the language of the agreement (i.e., that projects having 65 or more units of housing must be signatory to the PLA). These projects were identified using the PLA status given on the CA DIR website. These projects are

- FLOR 401 Lofts (Skid Row Housing Trust, 99 housing units)
- People Assisting the Homeless (PATH) Metro Villas (PATH Ventures, 122 housing units).

The reason for these projects being "noncompliant" with the PLA threshold appears to be the timing of the funding commitments relative to the passage of the PLA. These projects were the only proposed developments with 65 or more units that received funding commitments in the "pre-2017" round of HHH funding commitments, which took place before the drafting and adoption of the PLA.

In the descriptive statistics in Table A.1, I exclude these two projects so that they do not contribute to the average costs of projects that were subject to the PLA. This exclusion does not qualitatively influence the cost estimates discussed. Its effect on construction costs is demonstrated in Table A.2. Panel A replicates the construction costs from Table A.1 and panel B shows average construction costs including these two projects. The data points highlighted in gray indicate where estimates differ based on this exclusion. The inclusion of the two non-PLA projects above the threshold lowers the per unit costs of the 95+ units group by approximately $8,500 and lowers overall per unit construction costs by around $2,000. This is consistent with the notion that the PLA increased construction costs.

Table A.2. Construction Costs of Projects Using Alternate PLA-Related Sample Inclusion Criteria

	All	Fewer Than 50 Units	50 to 64 Units	65 to 94 Units	95 or More Units
Panel A. HHH Projects with Non-PLA 65+ Unit Projects Excluded					
Construction cost	341,277	357,273	328,798	374,033	325,427
	[55,800]	[55,507]	[40,623]	[42.523]	[83,649]
Observations	68	19	31	8	11
Panel B. HHH Projects with Non-PLA 65+ Unit Projects Included					
Construction cost	339,328	357,273	328,798	374,033	316,872
	[55,881]	[55,507]	[40,623]	[42.523]	[79,171]
Observations	71	19	31	8	13

SOURCE: Author calculations from city of Los Angeles, TCAC, and CDLAC data.
NOTE: Standard deviations for each measure in square brackets.

In the regression results in Chapter Five, I code these projects according to their actual PLA status rather than their "assigned" status so that the two noncompliant HHH projects above the PLA threshold contribute to the slope estimate for projects under 65 units (providing support for the cost slope for these "untreated" projects at two points above the PLA threshold, where there would otherwise be no data). The regression results do not change in a qualitatively meaningful way whether these projects are implemented as described above, included as if they "complied" with their assigned PLA status (so that they instead contribute to the slope estimate for PLA projects of 65 or more units), or are excluded from the estimates altogether. Table B.4 provides a comparison of estimates using specification 2 (described in Chapter Five) with these three ways of handling these three data points (assigning actual PLA status, assigning expected PLA status, or exclusion). In neither alternative case is the estimate affected in an economically or statistically meaningful way.

Appendix B: Regression Model Details and Sensitivity Analyses

Statistical Model Description

The statistical model used to conduct the cost analyses for this study takes the following form:

$$y_{it} = \alpha_i + \beta_0 HHH_i + \beta_1 65units_i + \beta_2 HHHPLA_i + \gamma_0 units_i + \gamma_1(units_i \times PLA_i) + X_i'\Phi + \delta_t + \epsilon_{it}.$$

This model regresses the per unit construction cost, y—measured either in \$1,000s or the natural log of \$1,000s—for project i estimated in year t on a binary indicator variable for using HHH funding, a binary indicator variable for the project comprising 65 units or more of housing, and the interaction between these two indicator variables, which is renamed, simply, *HHHPLA*. Also included is a measure of the number of housing units that has been centered around zero (i.e., a running variable), and this same variable interacted with an indicator for having 65 or more housing units (allowing estimation of a discrete slope for projects above the PLA threshold). A number of controls, detailed below, are included in the vector X, and a set of year (of LIHTC application) dummy variables, δ_t, are included to control for otherwise unobservable common shocks to construction costs. The coefficient of interest is β_2, which captures the association between per unit construction costs and being subject to the HHH PLA.

A Discussion of Model Identification

My preferred interpretation of this model is as a cross-sectional, two-way fixed effects model of the type commonly used in DD estimation that controls linearly for economies of scale in construction above and below the PLA threshold through the use of running variable-type cost slopes, as described above. Considering this model in the simple DD paradigm of two groups, one treated and one untreated, and two states, pretreatment and posttreatment, the treated group is the sample of HHH projects, the posttreatment state is the sample of projects of 65 or more units, and the object of interest is the interaction of these two conditions. Viewed through this lens, the identifying assumption is that, in the absence of the PLA, the distribution of project sizes would have been the same for the HHH-funded projects as for the non–HHH-funded projects. Under this simpler assumption, any cost discontinuity at the PLA threshold is attributable to the PLA. The analysis in the report suggests this is a conceptually and empirically plausible assumption. Here the data exclusions around the PLA threshold (specification 3 and the other specifications in Table B.6) can be thought of as a sensitivity test (specifically, sensitivity of the economies of scale controls and, thus, the discontinuity between them to excluding data where the policy of interest may affect choice of project size).

However, it may also be instructive to consider how any bias might bear on the validity of the estimates under a DRD interpretation. Conceptually, this manipulation could be upward or downward in magnitude. First, it may be that developers unable to pursue a cost-effective strategy to developing a larger project under the PLA might choose a project size under the threshold while developers able to operate effectively under the PLA might choose to build a larger project that is subject to the PLA. This would result in an estimate that would be biased downward in magnitude (toward zero). On the other hand, it might also be the case that developers who were more aware of the presence and implications of the PLA chose to build below the threshold, while a smaller number of those who were not aware of the PLA chose to build larger projects subject to it. If such developers tended to be less efficient, then this could bias the estimate effect upward.

The data exclusions implemented in specification 3 and in Table B.6 are a now-common method of addressing this type of bias often referred to as the "donut hole" approach (Barreca, Lindo, and Waddell, 2016; Cattaneo, Idrobo, and Titiunik, 2019). The results indicate that excluding projects between 64 and 74 units has the effect of slightly increasing the size of the estimate (by around 8 percent), while the additional (broader) exclusions in columns 2 and 3 of Table B.6 bracket this larger estimate, with the results in column 2 around 20 percent larger than the preferred specification (2) in the main results and the results in column 3 of Table B.6 nearly identical to this same estimate. Thus, any bias from this type of sorting around the PLA threshold appears to be modestly downward in magnitude, but sensitivity of the estimates to various-sized data exclusions appears to be bracketed by an increase in the magnitude of the estimate of between 0 and 20 percent.

Control Variables

Below is a list of control variables included in the model and a brief summary of the motivation that underlies the inclusion of each:

- *Shares of unit type in each project* (i.e., proportion of units that are studio, one bedroom, two bedroom, or three or more bedrooms). The motivation for including this variable is that a project comprising all three-bedroom units will necessarily have a higher per unit cost than a project of all studio units since the former units are simply larger, meaning that each unit costs more in materials. The inclusion of this variable along with the number of units in each project also effectively controls for the overall size of the project since the actual sizes of units of different types are typically quite close to the minimum unit sizes specified by the city. Thus, the number of units times the share of units of different types provides a fairly accurate relative measure of overall project size.[32]

---

[32] Analyzing a subsample of 16 projects that have approved plans on file with the city of Los Angeles Department of Planning, studio units averaged 110 percent of the minimum size and 88 percent of the maximum size specified

- *Share of units that are supportive housing.* The number of supportive housing units is correlated with potential cost drivers such as the amount of common/service areas (e.g., meeting rooms, case worker offices) required, so controlling for the share of these units is important.

- *Stories/construction type.* The number of stories can influence project size by directly increasing costs. As stories increase, different building types must be used (e.g., wood over a concrete foundation versus steel construction). However, it can also increase economies of scale (e.g., building additional stories over a common building foundation with shared electrical and plumbing infrastructure can be less costly than building multiple single-story buildings). To control for these forces, I use, first, a linear measure of stories as a continuous variable and, second, a pair of indicator variables for groups of stories that are likely to be strongly correlated with different building types. The first of these is an indicator for six- to eight-story buildings, which are most likely to be wood-over-podium-type construction, and a second indicator for buildings of nine or more stories, which are likely to be steel and concrete (type I) construction.

- *Prevailing wage and commercial prevailing wage* (PW/CPW). A pair of indicator variables control for the requirement to pay one of two tiers of prevailing wages in California: residential prevailing wage and a higher commercial prevailing wage scale. All HHH-funded projects are required to pay at least residential prevailing wages. Not all non–HHH-funded projects are required to do so (specifically, six of these projects did not indicate such a requirement). The requirement to pay CPW is identified here as applying to projects of five or more stories, as specified by both the state of California's DIR website and the city of Los Angeles's requirements for Proposition HHH funding (see section 3.7.6 of Proposition HHH Permanent Supportive Housing: Program Regulations, Policies, and Procedures (City of Los Angeles, 2020). This variable is implemented as a binary indicator for projects of five or more stories. I note that this variable may understate the number of projects subject to the commercial prevailing wage since aspects of other structures that are part of an overall project may also trigger this requirement.

- *Elevator/parking.* These two variables are indicator variables representing data from LIHTC funding applications. TCAC/CDLAC, the agencies that allocate this funding, allow for higher costs for properties where at least 95 percent of the project's upper-floor units are serviced by an elevator. Similarly, the parking variable indicates whether the project involves either subterranean parking or an additional parking structure, both of which trigger an allowance for higher project costs.[33]

---

by the city's Housing and Community Investment Department, while one-bedroom units averaged 119 percent of the minimum size and 85 percent of the maximum size.

[33] The coefficients on an elevator are large and negative, suggesting that this measure may be serving as a proxy for economies of scale inherent in larger buildings, which are more likely to have this amenity. For this reason, I do not focus on the specific magnitude of this (and other coefficients) per se, but consider whether (a) they improve the overall fit of the model and (b) they improve the precision of the estimated PLA effect. The inclusion of this variable meets both of these criteria. I also note that the latter parking requirements may also be correlated with commercial prevailing wage requirements. For this reason, I do not focus on the results of the model for commercial prevailing wage per se. Rather, I assume that the stories-based indicator variable along with this parking variable likely

- *Transit-oriented development (TOD).* This is an indicator variable equal to 1 for projects sited within a half mile of a major transit stop. This qualification results in a reduction of 50 percent in the HHH interest rate and also affects, for example, the amount of LIHTC credits that can be applied for (a project's "eligible basis"), both of which may affect a project's costs and financial viability.
- *Controls for a project's target population type(s).* Specific characteristics of projects, such as the size and makeup of common areas or areas for the provision of services as well as other aspects of construction (e.g., accessibility accommodations), may be influenced by needs specific to various target groups of a given project. This is implemented as a set of three mutually exclusive binary controls for three key subpopulations:
  - individuals (omitted from the model, serving as the reference category)
  - families
  - special populations (e.g., veterans, survivors of domestic abuse or sexual trafficking, individuals suffering from mental illness, seniors, transition-aged youth, and HIV-positive individuals).

In some projects, these populations overlap, so they are implemented in a hierarchical fashion that prioritizes coding family projects (since these may require larger amounts of common area for meeting the service needs of both children and adults), special populations, and then individuals.

---

together control effectively for this potential cost driver in a way that allows for identification of the PLA's effect on costs, which is the estimate of interest.

Full Regression Result Tables

Table B.1. Estimates of Effect of PLA on Construction Costs

	(1) Full Sample	(2) Exclude Outlier Projects by Size	(3) Also Exclude 64- to 74-Unit Projects
65+ units	−28.523	−0.143	−4.482
	(24.412)	(25.922)	(30.889)
HHH project	24.897	25.032	20.639
	(22.642)	(20.574)	(23.135)
HHH PLA	**44.588[+]**	**43.344[*]**	**46.843[*]**
	(22.797)	**(20.914)**	**(22.572)**
Units	−0.127	0.047	0.483
	(0.552)	(0.608)	(0.831)
Units above 65	−0.774	−2.140[*]	−2.877[*]
	(0.648)	(0.868)	(1.147)
Stories	15.057	11.544	6.380
	(10.540)	(9.080)	(10.299)
6–8 stories	−20.059	−12.203	3.692
	(23.584)	(20.622)	(24.763)
9+ stories	−92.530	0.000	0.000
	(152.057)	(.)	(.)
Commercial PW	−4.596	4.288	16.434
	(15.063)	(13.441)	(15.874)
Residential PW	46.532[*]	25.868	19.642
	(22.606)	(19.854)	(20.840)
Share studio	−34.245[*]	−30.917[*]	−23.620
	(17.062)	(15.174)	(17.685)
Share 2 BR	−8.159	19.979	23.687
	(40.486)	(37.336)	(41.739)
Share 3+ BR	127.364[+]	102.501	105.069
	(65.439)	(61.997)	(70.771)

	(1) Full Sample	(2) Exclude Outlier Projects by Size	(3) Also Exclude 64- to 74-Unit Projects
Share PSH	−71.785*	−45.788[+]	−47.551[+]
	(26.128)	(25.036)	(27.555)
TOD	−26.629[+]	−31.752*	−27.276*
	(13.746)	(12.210)	(12.852)
Elevators	−37.043*	−44.961*	−48.031*
	(17.359)	(15.661)	(16.282)
Parking	7.206	11.378	11.760
	(11.236)	(10.042)	(11.238)
Families	−1.565	6.541	17.713
	(14.789)	(12.862)	(15.041)
Special populations	−9.056	−11.566	−14.194
	(11.732)	(10.292)	(11.121)
Constant	259.381*	271.125*	305.099*
	(54.233)	(48.279)	(51.704)
Observations	97	86	74
Adjusted R^2	0.636	0.735	0.759

NOTES: Dependent variable is the construction cost in $1,000s. Two projects with a CA Department of Industrial Relations PLA status that does not follow the 65+ threshold rule are omitted (see Appendix A for details). In specification 2, outlier projects are those below the 5th percentile or above the 95th percentile of project size. Standard errors in parentheses.
[+] $p < 0.10$, * $p < 0.05$.

Table B.2. Estimates of Effect of PLA on Construction Costs (Logged Dependent Variable)

	(1) Full Sample	(2) Exclude Outlier Projects by Size	(3) Also Exclude 64- to 74-Unit Projects
65+ units	−0.158*	−0.092	−0.118
	(0.079)	(0.078)	(0.092)
HHH project	0.054	0.039	0.039
	(0.073)	(0.062)	(0.069)
HHH PLA	0.195*	0.206*	0.217*
	(0.074)	(0.063)	(0.067)
Units	−0.000	0.000	0.002
	(0.002)	(0.002)	(0.002)
Units above 65	−0.003	−0.007*	−0.008*
	(0.002)	(0.003)	(0.003)
Stories	0.071*	0.054+	0.045
	(0.034)	(0.027)	(0.031)
6–8 stories	−0.098	−0.063	−0.027
	(0.076)	(0.062)	(0.074)
9+ stories	−0.606	0.000	0.000
	(0.491)	(.)	(.)
Commercial PW	−0.034	0.003	0.029
	(0.049)	(0.040)	(0.047)
Residential PW	0.175*	0.103+	0.080
	(0.073)	(0.060)	(0.062)
Share studio	−0.100+	−0.090+	−0.079
	(0.055)	(0.046)	(0.053)
Share 2 BR	−0.034	0.104	0.088
	(0.131)	(0.112)	(0.125)
Share 3+ BR	0.424*	0.279	0.293
	(0.211)	(0.187)	(0.211)
Share PSH	−0.265*	−0.168*	−0.168*
	(0.084)	(0.075)	(0.082)
TOD	−0.098*	−0.117*	−0.108*
	(0.044)	(0.037)	(0.038)

	(1) Full Sample	(2) Exclude Outlier Projects by Size	(3) Also Exclude 64- to 74-Unit Projects
Elevators	−0.123*	−0.146*	−0.155*
	(0.056)	(0.047)	(0.049)
Parking	0.014	0.026	0.025
	(0.036)	(0.030)	(0.034)
Families	-0.009	0.014	0.046
	(0.048)	(0.039)	(0.045)
Special populations	−0.030	−0.042	−0.055
	(0.038)	(0.031)	(0.033)
Constant	5.471*	5.514*	5.599*
	(0.175)	(0.145)	(0.154)
Observations	97	86	74
Adjusted R^2	0.673	0.792	0.814

NOTES: Dependent variable is log of construction cost in $1,000s. Two projects with a CA Department of Industrial Relations PLA status that does not follow the 65+ threshold rule are omitted (see Appendix A for details). In specification 2, outlier projects are those below the 5th percentile or above the 95th percentile of project size. Standard errors in parentheses.
[+] $p < 0.10$, [*] $p < 0.05$.

Sensitivity Testing Results and Regression Variable Descriptive Statistics

Table B.3. Estimates of Effect of PLA on Soft Costs

	(1) Full Sample	(2) Exclude Outlier Projects by Size	(3) Also Exclude 64- to 74-Unit Projects
65+ units	−28.785	1.141	8.070
	(18.534)	(22.857)	(24.842)
HHH project	−22.384	−19.726	−12.723
	(17.190)	(18.141)	(18.606)
HHH PLA	38.572*	37.682*	27.113
	(17.307)	(18.441)	(18.153)
Units	−0.634	−0.929[+]	−1.679*
	(0.419)	(0.536)	(0.669)
Units above 65	−0.234	−1.026	0.114
	(0.492)	(0.765)	(0.922)
Stories	2.314	0.045	−4.770
	(8.002)	(8.006)	(8.283)
6–8 stories	12.539	12.312	37.569[+]
	(17.905)	(18.184)	(19.916)
9+ stories	43.058	0.000	0.000
	(115.441)	(.)	(.)
Commercial PW	−8.370	−5.924	−12.861
	(11.435)	(11.851)	(12.767)
Residential PW	24.435	10.431	4.735
	(17.163)	(17.507)	(16.761)
Share studio	−3.720	0.030	6.878
	(12.954)	(13.380)	(14.223)
Share 2 BR	46.786	59.144[+]	65.171[+]
	(30.737)	(32.921)	(33.569)
Share 3+ BR	−48.607	−91.470[+]	−100.479[+]
	(49.681)	(54.666)	(56.918)
Share PSH	−39.885*	−45.309*	−26.946
	(19.836)	(22.075)	(22.161)

	(1) Full Sample	(2) Exclude Outlier Projects by Size	(3) Also Exclude 64- to 74-Unit Projects
TOD	−2.405	−6.890	−5.529
	(10.436)	(10.766)	(10.336)
Elevators	−12.937	−10.805	−5.367
	(13.179)	(13.809)	(13.094)
Parking	−8.048	−10.257	−24.541*
	(8.531)	(8.855)	(9.038)
Families	6.765	10.051	26.153*
	(11.228)	(11.341)	(12.097)
Special populations	14.677	16.039+	9.563
	(8.907)	(9.075)	(8.944)
Constant	157.976*	173.204*	168.012*
	(41.174)	(42.570)	(41.583)
Observations	97	86	74
Adjusted R^2	0.422	0.475	0.574

NOTES: Estimates are in $1,000s. Two projects with a CA Department of Industrial Relations PLA status that does not follow the 65+ threshold rule are omitted (see Appendix A for details). In specification 2, outlier projects are those below the 5th percentile or above the 95th percentile of project size. Standard errors in parentheses.
+ $p < 0.10$, * $p < 0.05$.

Table B.4. Estimated PLA Cost Effect Using Alternate PLA-Related Sample Inclusion Criteria

	Exclude Non-PLA 65+ Unit Projects	Include Non-PLA 65+ Projects and Use Expected PLA Status	Include Non-PLA 65+ Projects and Use Actual PLA Status
HHH PLA	44.588[+]	44.429[+]	42.781[+]
	(22.797)	(22.475)	(22.307)
	[0.054]	[0.052]	[0.059]
Observations	97	99	99
Adjusted R^2	0.636	0.642	0.641

NOTES: Second and third specifications above are as discussed in Appendix A. Estimates (in $1,000s) are generated using preferred model as described in Chapter Five and Appendix B text. Reported p-values are from a two-tailed t-test. Standard errors are in parentheses and p-values are in square brackets.
[+] $p < 0.10$, [*] $p < 0.05$ [**] $p < 0.01$.

Table B.5. Estimated PLA Cost Effect Using Quadratic Modeling of Unit Size

	Single Quadratic Specification	Independent Quadratic Terms Above/Below Threshold
HHH PLA	44.877[*]	45.457[*]
	(21.017)	(21.424)
	[0.037]	[0.038]
Observations	86	86
Adjusted R^2	0.735	0.728

NOTES: These model specifications add a quadratic term in unit size. In the first specification, a single quadratic term is used instead of the two independent linear terms specified in the main model. In the second specification, an independent quadratic term is estimated for projects above the PLA threshold. Estimates (in $1,000s) are generated using preferred model as described in Chapter Five and Appendix B text. Reported p-values are from a two-tailed t-test. Standard errors are in parentheses and p-values are in square brackets.
[+] $p < 0.10$, [*] $p < 0.05$ [**] $p < 0.01$.

Table B.6. Estimated PLA Cost Effects Using Alternate Data Exclusions Around the PLA Threshold

	Exclude Projects Between 64 and 74 Units	Exclude Projects Between 64 and 80 Units	Exclude Projects Between 60 and 90 Units
HHH PLA	46.843*	52.591*	44.646+
	(22.572)	(23.438)	(24.866)
	[0.043]	[0.030]	[0.081]
Observations	74	68	59
Adjusted R²	0.759	0.754	0.781

NOTES: Estimates (in $1,000s) are generated using specification 3 as described in Chapter Five and Appendix B text while varying bandwidth of excluded data as indicated in each column. Estimates (in $1,000s) are generated using preferred model as described in Chapter Five and Appendix B text. Reported p-values are from a two-tailed t-test. Standard errors are in parentheses and p-values are in square brackets.
$^+ p < 0.10,$ $^* p < 0.05$ $^{**} p < 0.01$.

Table B.7. Descriptive Statistics for Cost-Analysis Sample

	HHH Projects	Non-HHH Projects
Units	66.30	66.25
	(31.11)	(28.96)
PLA	0.275	0
	(0.450)	(0)
Stories	5.203	4.679
	(2.610)	(1.389)
6–8 stories	0.203	0.286
	(0.405)	(0.460)
9+ stories	0.0290	0
	(0.169)	(0)
Commercial PW	0.594	0
	(0.495)	(0)
Residential PW	1	0.786
	(0)	(0.418)
Share studio	0.484	0.272
	(0.389)	(0.359)
Share 2 BR	0.129	0.147
	(0.175)	(0.176)

	HHH Projects	Non-HHH Projects
Share 3 BR+	0.0543	0.105
	(0.0917)	(0.147)
Share PSH	0.801	0.677
	(0.223)	(0.304)
TOD	0.841	0.786
	(0.369)	(0.418)
Elevators	0.899	0.929
	(0.304)	(0.262)
Parking structure	0.565	0.750
	(0.499)	(0.441)
Families	0.290	0.250
	(0.457)	(0.441)
Special populations	0.406	0.536
	(0.495)	(0.508)
Observations	69	28

SOURCE: Author calculations from city of Los Angeles, TCAC, and CDLAC data.
NOTE: Standard deviations in parentheses.

Appendix C: Additional Figures and Tables

Figure C.1. Estimating the Effect of the PLA on Construction Costs While Incorporating Economies of Scale (All Data Points)

Panel A. Single Slope Cost Estimate and No Allowance for a Discontinuity at 65 Units

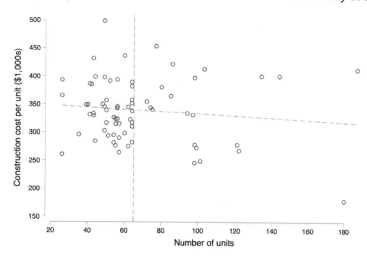

Panel B. Discrete Slope Estimates Above and Below 65 Units

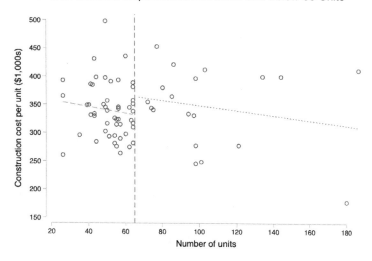

SOURCE: Author calculations from city of Los Angeles, TCAC, and CDLAC data.

Table C.1. Summary of Past Studies of Targeted Hiring Provisions with PLA Status

Jurisdiction	PLA	Mandatory	Good Faith	Compliance Monitoring	Goal		Outcome		
					Overall	Apprentice	Overall	Apprentice	Criteria
City of Cleveland		X	X		20 (hours)		31		City residents
City of Oakland			X	X	50 (hours)	15	30	11	City residents
City of Richmond			X	X	25 (hours)		**27**		City residents
West Contra Costa USD			X	X	24 (hours)	20	**58**	13	Priority characteristics
City College of San Francisco	X		X		40 (workers)	No goal	15	N/A	City residents
Oakland USD	X		X	X	50 (hours)	20	25	17	City residents w/priority tiers
LA Department of Public Works	X		X	X	30-40 (workers)	No goal	26	18	Zip code tiers
LAUSD*	X		X	X	50 (workers)	30	38	**31**	City residents w/zip priority
LA Community College District**	X		X	X	30 (hours)	30	**32**	No data	City residents w/zip priority
Community Redevelopment Agency of the City of Los Angeles	X		X	X	30 (hours)	30-40	11-47	No data	City residents w/zip priority
Port of Oakland	X		X	X	50 (hours)	20	**59**	9	City residents w/priority tiers
Cleveland University Hospital	X		X	X	20 (workers)	No goal	**reportedly met**	N/A	City residents

| | | | | | Goal | | Outcome | | |
Jurisdiction	PLA	Mandatory	Good Faith	Compliance Monitoring	Overall	Apprentice	Overall	Apprentice	Criteria
Washington, D.C. (baseball park)***	X		X	X	50 (journey workers)	50 (hours)	*26*	**70**	City residents

SOURCES: Luster et al. (2010); Figueroa, Grabelsky, and Lamare (2011); Owens-Wilson (2010). Projects from these reports with sufficiently comparable goals were used in this table. For ease of interpretation, goals that were met are in bold, and goals that were not met are in italics.
NOTES:
* The LAUSD PLA had an additional provision that 40 percent of apprentices should be first-year apprentices; the outcome was 31 percent.
** The LA Community College District PLA had an additional goal that 20 percent of local residents should include at-risk workers. The outcome was 9 percent.
*** The Washington, D.C., project had additional goals including that 50 percent of new hires overall were residents (exactly met), that 100 percent of new apprentice hires should be residents (outcome of 85 percent), and that 25 percent of total craft hours should be worked by apprentices (outcome 19 percent).

Table C.2. Developers of HHH and Non-HHH Projects in the Analysis Data

Developer	HHH Projects	Non-HHH Projects
Concerned Citizens Community Involvement	1	
1010 Development Corporation	1	1
A Community of Friends	3	3
Abbey Road, Inc.	3	
Abode Communities	4	
Affirmed Housing Group, Inc.	6	1
AMCAL Multi-Housing, Inc.		1
American Family Housing	1	
Azure Development, Inc.	1	
BRIDGE Housing Corporation	2	
Century Affordable Development, Inc.	1	
Chelsea Investment Corporation	2	
Clifford Beers Housing, Inc.	2	1
Coalition for Responsible Community Development	3	
Community Development Partners	1	
D. L. Horn & Associates, LLC		2
Daylight Community Development	2	
DDCM Incorporated	1	
Decro Corporation	4	
Deep Green Housing and Community Development	1	1
Domus GP, LLC	4	
EAH Housing	3	
East LA Community Corporation	1	2
Flexible PSH Solutions	2	
Gospel Truth CDC, Inc.	1	
GTM Holdings, LLC	1	1
Hollywood Community Housing Corporation		1
Highridge Costa Development Company, LLC	3	
Hollywood Community Housing Corporation	1	
Hope Street Development Group, LLC	2	
Innovative Housing Opportunities, Inc.	3	
John Stanley, Inc.	2	
Koreatown Youth & Community Center	2	
LA Family Housing Corporation	2	1
LINC Housing Corporation	3	1
Los Angeles Housing Partnership	1	
Mercy Housing California		1
Meta Housing Corporation	3	8

Developer	HHH Projects	Non-HHH Projects
Many Mansions	2	
Mercy Housing California	2	
The Pacific Companies	1	
PATH Ventures	2	2
The Richman Group of California	1	
Single Room Occupancy Housing Corporation	2	2
The Skid Row Housing Trust	4	1
Thomas Safran & Associates Development, Inc.	6	
Unique Construction & Development, Inc.	1	
Venice Community Housing Corporation	2	
Wakeland Housing and Development Corporation	4	
Weingart Center Association	2	
Western Community Housing, Inc.	1	
West Hollywood Community Housing Corporation	2	2
Women Organizing Resources, Knowledge and Services	2	1

SOURCE: Author calculations from city of Los Angeles and TCAC/CDLAC data.
NOTE: Some listed developers are jointly involved in certain projects. For this tabulation, involvement in a project, whether as the sole developer or a partner, is counted as a project. Three projects, the sole developer of which appeared to be a special-purpose entity related to developing one HHH project, were removed from this tabulation.

Abbreviations

AH	affordable housing
AH/PSH	affordable/permanent supportive housing
AH/SH	affordable/supportive housing
BR	bedroom
CA	California
CDLAC	California Debt Limit Allocation Committee
DD	difference-in-differences
DIR	Department of Industrial Relations
DOL	Department of Labor
DRD	difference in regression discontinuities
GAO	Government Accounting Office
HHH	Proposition HHH
HIV	human immunodeficiency virus
JJJ	Proposition JJJ
LA	Los Angeles
LAUSD	Los Angeles Unified School District
LIHTC	low-income housing tax credit
PLA	project labor agreement
PSH	permanent supportive housing
PW	prevailing wage
RD	regression discontinuity
TCAC	California Tax Credit Allocation Committee
THP	targeted hiring provision
TOD	transit-oriented development
USD	unified school district

References

Alpert Reyes, Emily, "L.A. Will Eliminate 'Veto' Provision for Homeless and Affordable Housing to Keep State Funding," *Los Angeles Times*, October 16, 2018. As of May 4, 2021: https://www.latimes.com/local/lanow/la-me-ln-homeless-letter-20181017-story.html

Associated Builders and Contractors, Inc., "Analysis of Bids and Costs to the Taxpayer for the Roswell Park Cancer Institute, New York State Dormitory Authority Construction Project," 1995. As of May 4, 2021: http://thetruthaboutplas.com/wp-content/uploads/2012/12/Roswell-Park-Cancer-Institute-NY-Letters-on-PLA-Cost-Increases-1995.pdf

Bachman, Paul, William F. Burke, and David G. Tuerck, *Policy Brief: The Anticompetitive Effects of Government-Mandated Project Labor Agreements on Construction in Washington State*, Seattle: Washington Policy Center, 2019. As of May 4, 2021: https://www.washingtonpolicy.org/library/doclib/Shannon-The-Anticompetitive-Effects-of-Project-Labor-Agreements-on-Construction-in-Washington-State.pdf

Bachman, Paul, and Jonathan Haughton, "Do Project Labor Agreements Raise Construction Costs?" *Case Studies in Business, Industry and Government Statistics*, Vol. 1, No. 1, 2007, pp. 71–79.

Bachman, Paul, Jonathan Haughton, and David G. Tuerck, *Project Labor Agreements and the Cost of Public School Construction in Connecticut*, Boston: Beacon Hill Institute at Suffolk University, 2004. As of May 4, 2021: https://beaconhill.org/BHIStudies/PLA2004/PLAinCT23Nov2004.pdf

Bachman, Paul, and David G. Tuerck, *Project Labor Agreements and Public Construction Costs in New York State*, Boston: Beacon Hill Institute at Suffolk University, 2006. As of May 4, 2021: https://thetruthaboutplas.com/wp-content/uploads/2012/12/PLA-and-Public-Construction-Costs-in-NY-State-BHI-2006.pdf

Barreca, Alan I., Jason M. Lindo, and Glen R. Waddell, "Heaping-Induced Bias in Regression-Discontinuity Designs," *Economic Inquiry*, Vol. 54, No. 1, 2016, pp. 268–293.

Belman, Dale, Russell Ormiston, Richard Kelso, William Schriver, and Kenneth A. Frank, "Project Labor Agreements' Effect on School Construction Costs in Massachusetts," *Industrial Relations: A Journal of Economy and Society*, Vol. 49, No. 1, 2010, pp. 44–60.

Brothers, Keith, "Connecticut Benefits from Project Labor Agreements," *The CT Mirror*, December 31, 2020. As of May 20, 2021:
https://ctmirror.org/category/ct-viewpoints/connecticut-benefits-from-project-labor-agreements/

Brubeck, Ben, "America Can Build Back Better Through Fair and Open Competition," *The Hill*, April 28, 2021. As of May 20, 2021:
https://thehill.com/opinion/finance/550792-america-can-build-back-better-through-fair-and-open-competition

Byrne, Thomas H., Benjamin F. Henwood, and Anthony W. Orlando, "A Rising Tide Drowns Unstable Boats: How Inequality Creates Homelessness," *Annals of the American Academy of Political and Social Science*, Vol. 693, No. 1, 2021, pp. 28–45.

Calandro, Alan, "Speak the Truth About Project Labor Agreements," *The CT Mirror*, January 7, 2021. As of May 20, 2021:
https://ctmirror.org/category/ct-viewpoints/speak-the-truth-about-project-labor-agreements/

California Code of Regulations, Title 4, Division 15, Chapter 1, Federal and State Low-Income Housing Tax Credit, 2020. As of May 21, 2021:
https://govt.westlaw.com/calregs/Browse/Home/California/CaliforniaCodeofRegulations?guid=I47630370D45A11DEA95CA4428EC25FA0&originationContext=documenttoc&transitionType=Default&contextData=(sc.Default)

California Department of Housing and Community Development, "Multifamily Housing Program, July 2020, Notice of Funding Availability," State of California memorandum, July 15, 2020.

California Department of Housing and Community Development, California Tax Credit Allocation Committee, California Housing Finance Agency, and California Debt Limit Allocation Committee, *Affordable Housing Cost Study: Analysis of the Factors That Influence the Cost of Building Multi-Family Affordable Housing in California*, 2014.

California Department of Industrial Relations, "Public Search Utility Tool," 2010. As of June 1, 2021:
https://www.dir.ca.gov/pwc100ext/ExternalLookup.aspx

Cantrell, Jennifer D., and Suparna Jain, *Enforceability of Local Hire Preference Programs*, Vol. 59, National Cooperative Highway Research Program Legal Research Digest, Washington, D.C.: National Academy of Sciences, 2013.

Card, David, and Alan B. Krueger, "Minimum Wages and Employment: A Case Study of the Fast-Food Industry in New Jersey and Pennsylvania," *American Economic Review*, Vol. 84, No. 4, 1994, pp. 772–793.

Cattaneo, Matias D., Nicolás Idrobo, and Rocío Titiunik, *A Practical Introduction to Regression Discontinuity Designs*, Cambridge: Cambridge University Press, 2019.

Cedillo, Gilbert, Paul Krekorian, David Ryu, Marqueece Harris-Dawson, Herb J. Wesson, Jr., and Mitchell Englander, "Motion on Proposition HHH Bond Proceeds to Los Angeles City Council," May 3, 2017. As of May 12, 2021:
https://www.gilcedillo.com/5_5_17

Cellini, Stephanie Riegg, Fernando Ferreira, and Jesse Rothstein, "The Value of School Facility Investments: Evidence from a Dynamic Regression Discontinuity Design," *Quarterly Journal of Economics*, Vol. 125, No. 1, 2010, pp. 215–261.

Cervantes, Rushmore D., *Proposition HHH Permanent Supportive Housing Loan Program: Funding Recommendations 2018–19 Call for Projects Round 3, Memorandum to Proposition HHH Citizens Oversight Committee*, City of Los Angeles Housing + Community Investment Department, August 20, 2019.

City of Los Angeles, "Homelessness Reduction and Prevention, Housing, and Facilities Bond—Question & Answer," City Administrative Officer, undated. As of May 4, 2021:
https://cao.lacity.org/Homeless/Homelessness%20Bond%20QA.pdf

———, *Proposition HHH Permanent Supportive Housing: Program Regulations, Policies, and Procedures, draft*, Los Angeles: Bureau of Contract Administration, 2020.

———, *Proposition HHH Project Labor Agreement (PLA) with Los Angeles/Orange Counties Building and Construction Trades Council*, Los Angeles: Bureau of Contract Administration, 2018.

———, "Recommendation Report, Case CPC-2017-3136-CA, December 14, 2017," Department of City Planning, 2017. As of June 4, 2021:
http://planning.lacity.org/StaffRpt/InitialRpts/CPC-2017-3136.pdf

———, *Rules and Regulations: Implementing the First Source Hiring Ordinance*, Los Angeles: Bureau of Contract Administration, Office of Contract Compliance, 2016.

City of Los Angeles Department of Public Works, "Measure JJJ: Affordable Housing and Labor Standards Related to City Planning," Bureau of Contract Administration, 2021. As of May 20, 2021:
https://bca.lacity.org/measure-JJJ

City of Pasadena, "First Source Local Hiring," Department of Finance, 2021. As of June 4, 2021:
https://www.cityofpasadena.net/finance/doing-business-with-the-city/first-source-local-hiring

Conrad N. Hilton Foundation, "Indicators of Community Progress toward the Goal," 2018. As of May 4, 2021:
http://assets.hiltonfoundation.org/homelessnessreport/psh-units-indicators-of-community -progress-toward-the-goal.html

———, "Developing and Passing Proposition HHH and Measure H in Los Angeles," 2019. As of June 4, 2021:
https://www.hiltonfoundation.org/wp-content/uploads/2019/10/HHH___H_Final_One_Sheet -3.pdf

Duncan, Kevin, and Russell Ormiston, "What Does the Research Tell Us About Prevailing Wage Laws?" *Labor Studies Journal*, Vol. 44, No. 2, 2019, pp. 139–160.

Duong, Howard T., *Los Angeles Homelessness Epidemic: An Analysis of Proposition HHH*, San Luis Obispo: California Polytechnic State University, 2021. As of May 10, 2021:
https://digitalcommons.calpoly.edu/cmsp/432/

ENR California, "PLA Study on School Construction Costs Disputed," *Engineering News-Record California Views Blog*, August 12, 2011. As of May 20, 2021:
https://www.enr.com/blogs/12-california-views/post/13682-pla-study-on-school -construction-costs-disputed

Figueroa, Maria, Jeff Grabelsky, and Ryan Lamare, *Community Workforce Provisions in Project Labor Agreements: A Tool for Building Middle-Class Careers*, Ithaca, N.Y.: ILR School, Cornell University, 2011.

Fiore, Nichole, Rian Watt, Carol Wilkins, Brooke Spellman, and Rebecca Jackson, *Increasing and Accelerating the Development of Permanent Supportive Housing in Los Angeles*, Rockville, Md.: Abt Associates, 2019.

Flood, Sarah, Miriam King, Renae Rodgers, Steven Ruggles, and J. Robert Warren, *Integrated Public Use Microdata Series, Current Population Survey: Version 8.0 [dataset]*, Minneapolis, Minn.: IPUMS, 2020. As of August 11, 2021:
https://doi.org/10.18128/D030.V8.0

Galperin, Ron, *High Cost of Homeless Housing: Review of Proposition HHH*, Los Angeles: Controller's Office, City of Los Angeles, 2019. As of May 4, 2021:
https://lacontroller.org/audits-and-reports/high-cost-of-homeless-housing-hhh/

GAO—see General Accounting Office.

General Accounting Office, *Labor-Management Relations: Construction Agreement at DOE's Idaho Laboratory Needs Reassessing*, Washington, D.C.: Government Printing Office, GGD-91-80BR, 1991. As of June 2, 2021:
https://www.gao.gov/products/ggd-91-80br

Hernandez, Jennifer, "California Environmental Quality Act Lawsuits and California's Housing Crisis," *Hastings Environmental Law Journal*, Vol. 24, No. 1, 2018, pp. 21–71.

Holland, Gale, "L.A. Officials Launch Campaign for Homeless Housing Bond Measure," *Los Angeles Times*, September 12, 2016. As of May 4, 2021:
https://www.latimes.com/local/lanow/la-me-ln-homeless-bond-kickoff-20160912-snap
-story.html

Holland, Gale, and Doug Smith, "L.A. Votes to Spend $1.2 Billion to House the Homeless. Now Comes the Hard Part," *Los Angeles Times*, November 9, 2016. As of May 4, 2021:
https://www.latimes.com/local/lanow/la-me-ln-homeless-20161108-story.html

Koseff, Alexei, "California Housing Bills Run into Wall of Union Resistance," *San Francisco Chronicle*, August 25, 2020. As of June 3, 2021:
https://www.sfchronicle.com/politics/article/California-housing-bills-run-into-wall-of-union
-15514503.php

Lemieux, Thomas, and Kevin Milligan, "Incentive Effects of Social Assistance: A Regression Discontinuity Approach," *Journal of Econometrics*, Vol. 142, No. 2, 2008, pp. 807–828.

Los Angeles Community College District, "Amended Project Labor Agreement: Propositions A, AA, Measure J, and General Construction, Renovation and Rehabilitation Projects," 2001. As of May 20, 2021:
http://az776130.vo.msecnd.net/media/docs/default-source/labor-compliance/2020-pla-with
-amendments.pdf?sfvrsn=0

Los Angeles County Metropolitan Transportation Authority, "Project Labor Agreement, draft, ca. January 27," 2017. As of May 20, 2021:
https://metro.legistar.com/View.ashx?M=F&ID=4920513&GUID=4CB4B5DD-1189-4FFD
-A5CA-241C7A0FD3FA

Los Angeles Times Editorial Board, "Editorial: This Is Not the Time to Claw Back L.A. Homeless Housing Money to Build Shelters," *Los Angeles Times*, February 24, 2021. As of May 4, 2021:
https://www.latimes.com/opinion/story/2021-02-24/editorial-not-the-time-claw-back-la
-homeless-housing-money-to-build-shelters

Los Angeles Times Editorial Department, "Editorial: Debunking Eight Myths About Homelessness and Proposition HHH," *Los Angeles Times*, October 29, 2016. As of May 4, 2021:
https://www.latimes.com/opinion/editorials/la-ed-homelessness-myths-prop-hhh-20161028
-story.html

Los Angeles Unified School District, *Project Stabilization Agreement: New School Construction and Major Rehabilitation Funded by Proposition BB and/or Measure K*, Los Angeles: LAUSD Board of Education, 2003. As of May 20, 2021:
https://www.laschools.org/documents/download/project_stabilization_agreement%2fProject_Stabilization_Agreement_and_Side_Letters.pdf

Lund, John, and Joe Oswald, "Public Project Labor Agreements: Lessons Learned, New Directions," *Labor Studies Journal*, Vol. 26, No. 3, 2001, pp. 1–23.

Luster, Laura, Michael Potepan, Nadine Wilmot, with Michael Bernick, *Labor Market Analysis, San Francisco Construction Industry, Final Report*, Oakland, Calif.: L. Luster & Associates, 2010.

Mai-Duc, Christine, "California Needs More Affordable Homes. This Union Stands in the Way," *Wall Street Journal*, April 17, 2021. As of May 4, 2021:
https://www.wsj.com/articles/california-needs-more-affordable-homes-this-union-stands-in-the-way-11618660801

Martindale, Scott, "Union-Only O.C. Hiring Pacts Raise Alarms," *Orange County Register*, April 3, 2013. As of May 20, 2021:
https://www.ocregister.com/2013/04/03/union-only-oc-hiring-pacts-raise-alarms

McCrary, Justin, "Manipulation of the Running Variable in the Regression Discontinuity Design: A Density Test," *Journal of Econometrics,* Vol. 142, No. 2, 2008, pp. 698–714.

McShane, Blakeley B., David Gal, Andrew Gelman, Christian Robert, and Jennifer L. Tackett, "Abandon Statistical Significance," *American Statistician*, Vol. 73, No. supp., 2019, pp. 235–245.

Moran, John, "Pros and Cons of Using Project Labor Agreements," November 2, 2011. As of May 20, 2021:
https://www.cga.ct.gov/2011/rpt/2011-R-0360.htm

National Alliance to End Homelessness, "Permanent Supportive Housing," 2020. As of May 12, 2021:
https://endhomelessness.org/ending-homelessness/solutions/permanent-supportive-housing/

New Jersey Department of Labor, *Annual Report to the Governor and Legislature: Use of Project Labor Agreements in Public Works Building Projects in Fiscal Year 2008—as Required by the Project Labor Agreement (PLA) Act P.L. 2002, Chapter 44 (C.52:38-et seq.)*, Trenton: New Jersey Department of Labor and Workforce Development, 2010.

Northrup, Herbert R., and Linda E. Alario, "'Boston Harbor'–Type Project Labor Agreements in Construction: Nature, Rationales, and Legal Challenges," *Journal of Labor Research*, Vol. 19, No. 1, 1998, pp. 1–63.

Oreskes, Benjamin, "Councilman Kevin de León Wants 25,000 Housing Units for Homeless by 2025," *Los Angeles Times*, January 12, 2021. As of May 4, 2021:
https://www.latimes.com/homeless-housing/story/2021-01-12/councilman-kevin-de-leon-t
-wants-25-000-new-housing-units-for-homeless-people-by-2025

Oreskes, Benjamin, and David Zahniser, "L.A. Plans Nearly $1 Billion in Spending to Address Homelessness Under Garcetti Plan," *Los Angeles Times*, April 19, 2021. As of May 4, 2021:
https://www.latimes.com/homeless-housing/story/2021-04-19/los-angeles-will-increase
-budget-for-addressing-homelessness

Owens-Wilson, Sebrina, *Constructing Buildings and Building Careers: How Local Governments in Los Angeles Are Creating Real Career Pathways for Local Residents*, Oakland, Calif.: Partnership for Working Families, 2010.

Philips, Peter, and Scott Littlehale, *Did PLAs on LA Affordable Housing Projects Raise Construction Costs?* Salt Lake City: University of Utah, Department of Economics, 2015. As of May 4, 2021:
https://economics.utah.edu/research/publications/2015_03.pdf

Philips, Peter, and Emma Waitzman, *Project Labor Agreements and Bidding Outcomes: The Case of Community College Construction in California*, Berkeley, Calif.: UC Berkeley Labor Center, 2017. As of May 20, 2021:
https://laborcenter.berkeley.edu/project-labor-agreements-and-bidding-outcomes/

Raetz, Hayley, Teddy Forscher, Elizabeth Kneebone, and Carolina Reid, *The Hard Costs of Construction: Recent Trends in Labor and Materials Costs for Apartment Buildings in California*, Berkeley, Calif.: Terner Center for Housing Innovation at UC Berkeley, 2020.

Reamer, John L., "Consideration to the Joint Homelessness and Poverty and Housing Committees Regarding a Project Labor Agreement for Proposition HHH (Council File 17-0090-S1), Memorandum to the Honorable Members of the City Council," Bureau of Contract Administration, City of Los Angeles, 2017.

Reid, Carolina, *The Costs of Affordable Housing Production: Insights from California's 9% Low-Income Housing Tax Credit Program*, Berkeley, Calif.: Terner Center for Housing Innovation at UC Berkeley, 2020.

Ruggles, Steven, Sarah Flood, Sophia Foster, Ronald Goeken, Jose Pacas, Megan Schouweiler, and Matthew Sobek, *IPUMS USA: Version 11.0 [dataset]*, Minneapolis, Minn.: IPUMS, 2021. As of August 11, 2021:
https://doi.org/10.18128/D010.V11.0

Sewill, Ann, *Amendment to Proposition HHH FY 2020–2021 Project Expenditure Plan: Memorandum to the Proposition HHH Citizens Oversight Committee, February 19, 2021*, City of Los Angeles Housing + Community Investment Department, 2021.

Sharp, Steven, "Costs Rise for Measure HHH Supportive Housing Developments," 2020. As of May 4, 2021:
https://urbanize.city/la/post/costs-rise-measure-hhh-supportive-housing-developments

Smith, Dakota, "Lawsuits Target City over New Laws for Homeless Housing Projects and Motel Conversions," *Los Angeles Times*, May 13, 2018. As of May 4, 2021:
https://www.latimes.com/local/lanow/la-me-ln-lawsuit-homeless-lawsuit-20180511 -story.html

Smith, Doug, "Q&A: Proposition HHH Would Raise Funds to Build Homeless Housing in L.A.," *Los Angeles Times*, October 19, 2016. As of May 4, 2021:
https://www.latimes.com/local/california/la-me-ln-prop-hhh-qa-20161017-snap-story.html

State Policy Network, "About State Policy Network," 2021. As of May 20, 2021:
https://spn.org/state-policy-network-about/

Thistlethwaite, Donald L., and Donald T. Campbell, "Regression-Discontinuity Analysis: An Alternative to the Ex Post Facto Experiment," *Journal of Educational Psychology*, Vol. 51, No. 6, 1960, pp. 309–317.

Vasquez, Vince, Dale Glaser, and W. Erik Bruvold, *Measuring the Cost of Project Labor Agreements on School Construction in California*, La Jolla, Calif.: National University System Institute for Policy Research, 2010. As of May 12, 2021:
https://www.heartland.org/publications-resources/publications/measuring-the-cost-of-project -labor-agreements-on-school-construction-in-california

Ward, Jason, *Replication Code and Data for "The Effects of Project Labor Agreements on the Production of Affordable Housing: Evidence from Proposition HHH,"* Santa Monica: RAND Corporation, 2021. As of July 23, 2021:
https://github.com/RANDCorporation/HHH-Project-Labor-Agreements

Warth, Gary, "Gov. Newsom Proposes $12 Billion to House California's Homeless," *Los Angeles Times*, May 11, 2021. As of June 3, 2021:
https://www.latimes.com/california/story/2021-05-11/california-governor-proposes-12b-to -house-states-homeless

Woetzel, Jonathan, Tim Ward, Shannon Peloquin, Steve Kling, and Sucheta Arora, *Affordable Housing in Los Angeles: Delivering More—and Doing It Faster*, New York: McKinsey Global Institute, 2019. As of May 17, 2021:
https://www.mckinsey.com/industries/public-and-social-sector/our-insights/affordable- housing-in-los-angeles-delivering-more-and-doing-it-faster

Wooldridge, Jeffrey M., *Introductory Econometrics: A Modern Approach*, 7th ed., Independence, Ky.: Cengage, 2020.